Advance Praise

Energy Switch

Germany's phenomenal success with renewable energy offers a model that North Americans would be wise to emulate. In *Energy Switch*, Craig Morris tells us why Germany has become the world leader in this fast growing, multibillion dollar industry, and why America will be left behind again — if it doesn't act fast.

— Paul Gipe, advocate and critic, leader of the campaign for Advanced Renewable Tariffs in Ontario, Canada, and author of *Wind Energy Comes of Age* and *Wind Power: Renewable Energy for Home, Farm and Business*

Energy Switch is a well-informed and committed practical encouragement for the general change to renewable energies. It is a mission possible.

— Hermann Scheer, Recipient of the Alternative Nobel Prize, General Chairman of the World Council for Renewable Energy

Morris deals with the standard arguments against wind power so convincingly that you will want to have a whole book by him on this topic.

— *Photon Magazine*

Craig Morris has written the best "guide book" to renewable energy I've seen yet. It covers all the energies we should switch to, and he also deals with coal and nuclear, and why they not only can't save us, they might destroy us. Morris writes with the layman in mind; informative, not too technical, and not didactic. He tells us why Europe, and especially Germany, is so very far ahead of the US in using solar energy, mainly PV. Americans should wonder why — with *Energy Switch*, they can find out!

— Neville Williams, author of *Chasing The Sun: Solar Adventures Around The World*

ENERGY
SWITCH

ENERGY SWITCH

PROVEN SOLUTIONS
FOR A RENEWABLE FUTURE

CRAIG MORRIS

NEW SOCIETY PUBLISHERS

Cataloging in Publication Data:
A catalog record for this publication is available from the National Library of Canada.

Cover design by Diane MacIntosh. Photo: Brand X Pictures/Alamy

Printed in Canada.
First printing May 2006.

Paperback ISBN 13: 978-0-86571-559-2
Paperback ISBN 10: 0-86571-559-9

To order directly from the publishers, please call toll-free (North America) 1-800-567-6772, or order online at www.newsociety.com.
Any other inquiries can be directed by mail to:

New Society Publishers
P.O. Box 189, Gabriola Island,
BC V0R 1X0, Canada
1-800-567-6772

New Society Publishers' mission is to publish books that contribute in fundamental ways to building an ecologically sustainable and just society, and to do so with the least possible impact on the environment, in a manner that models this vision. We are committed to doing this not just through education, but through action. We are acting on our commitment to the world's remaining ancient forests by phasing out our paper supply from ancient forests worldwide. This book is one step toward ending global deforestation and climate change. It is printed on acid-free paper that is **100% old growth forest-free** (100% post-consumer recycled), processed chlorine free, and printed with vegetable-based, low-VOC inks. For further information, or to browse our full list of books and purchase securely, visit our website at: www.newsociety.com.

NEW SOCIETY PUBLISHERS www.newsociety.com

Contents

Acknowledgments

THIS BOOK WAS ORIGINALLY PUBLISHED IN GERMAN. It has not only been translated but also adapted to take account of the background knowledge of North American readers.

I would like to thank my German editor Florian Rötzer and my German publishing house Heise Verlag for their support; José Etcheverry of Canada's David Suzuki Foundation for his initial interest in my book; British-born author Guy Dauncey for the coaching; all of the experts and authors cited for their participation; the people at Austin Energy and Solar Austin, especially Anne Johnson, for hosting my lecture, in which I learned a lot about how to present this material to Americans; the New Orleans Alliance for Affordable Energy, especially Charles Reith and Karen Wimpelberg, for hosting several lectures before audiences that were not yet "converted"; the Delta Chapter of the Sierra Club, especially Micah Walker Parkin, for allowing me to present my material along with other experts to environmentalists; and the staff of New Society Publishers for their friendly assistance. Finally, thanks to my parents for giving a little boy from a swamp such good opportunities.

Any errors herein are solely mine.

Readers are invited to take part in a discussion about this book and its contents at <http://groups.yahoo.com/group/energyswitch>.

Instructions to Readers

THIS BOOK WAS ORIGINALLY PUBLISHED ONLINE from 2002-2004 as a series of articles in German. The Notes therefore generally contain links that readers of the online version could click on. Readers of this book may find it easier to simply search the Web for a title than to type in an entire URL, but the URLs may come in handy where search engines fail.

Readers are also advised to enter links no longer accessible into <www.archive.org>, where many websites are stored in previous versions. However, newspapers are removing articles from the Google cache and from archive.org so they can charge fees for their own digital archives. In cases where this has already happened, I have provided reference information for the print version available in libraries.

Chapter 1

Best Practices from Germany

In the meantime, too many Christian Democrats support renewables and wind power for there to be drastic clear-cutting of wind turbines.

— Jochen Ahn, chairman of Germany's ABO Wind AG, just before the German elections in September 2005 in which the conservative Christian Democrats defeated the governing coalition of Social Democrats and Greens

THIS BOOK DESCRIBES A PATH toward a sustainable energy supply. It is largely inspired by the success of European Union (EU) policy, and German policy in particular. Why should North American readers be interested? To begin with, Germany is the world's leader in wind energy and photovoltaics. And the EU is taking steps to increase the share of a broad range of renewables — including biomass, geothermal, and ocean power — in its energy supply over the next few decades. Obviously, if you are a proponent of renewables you'll be interested in what the EU is doing right. If you are not a proponent of renewables you will be especially interested in the limitations and drawbacks of coal, nuclear power, gas, and oil. This book does not divide sources of energy into "good guys and bad guys" but rather points out the necessity — indeed, the inevitability — of a switch to renewables. Though inevitable, this switch will be painful if we wait too long.

This book is also intended as a contribution to the discussions launched, on the one hand, by the 2004 essay by Michael Shellenberger and Ted Nordhaus titled "The Death of Environmentalism"[1] and, on the other hand, by Sharon Beder's 2003 book *Power Play: The Fight to Control the World's Electricity.*[2] Shellenberger and Nordhaus, themselves US environmentalists, criticize the US environmental community for overlooking the big picture. They write: "Environmentalists are learning all the wrong lessons from Europe. We closely scrutinize the *policies* without giving much thought to the *politics* that made the policies possible."

They then proceed to summarize the input from "25 of the environmental community's top leaders, thinkers and funders" — none of whom is European, though probably all of them are well aware of what Europe has been doing recently. Shellenberger and Nordhaus propose that the environmental community define its obstacles more broadly. I support this approach. But many readers perceived "The Death of Environmentalism" as blaming environmentalists for the obvious shortcomings of the green movement in the US. Shellenberger and Nordhaus complain that American environmentalists fail to articulate:

> a vision of the future commensurate with the magnitude of the crisis. Instead, they are promoting technical policy fixes like pollution controls and higher mileage standards — proposals that provide neither the popular inspiration nor the political alliances the community needs to deal with the problem.

But this description could apply equally to European environmentalists. I know of no work that sets out a unified "European vision" of the future (though the German-American collaboration *Factor Four*[3] comes close), but I have read plenty of European papers on topics such as new technologies, sustainable urban planning, and financing. So while Shellenberger and Nordhaus seem to suggest that Americans should follow the European example (an example they neglect to describe), in fact the main difference is not what environmentalists say and do in Europe but the opposition they face.

Almost all Europeans agree that the scarcity of natural resources threatens to destroy civilization, and they are undertaking some modest steps to prevent the worst. For example, Europeans are arguing not about whether we are running out of oil but about what the contribution of biodiesel, hybrids, fuel cells, and efficiency can be. They are not arguing not about whether industrialization is accelerating climate change but about whether nuclear power, carbon sequestration, a full switch to renewables along with conservation and efficiency, or a mixture of these is the best solution. Europe's success with renewables may be the result of a different attitude. The technology in Europe is the same as in North America, but in Europe there is less ideological opposition to renewables. It would be wrong to criticize US environmentalists for the way some of their opponents prevent informed, open debate in the US.

While political factors would be an excellent starting point, I would like to postpone this discussion. First we need to discuss the potential of various sources of energy, focusing all the while on whether the EU might be doing something worth copying, for it is my contention that Europeans are not better people but their relative lack of resources has forced them to lead the way toward renewables. The US soon will be forced to follow suit.

Beder's *Power Play* is a must-read for those interested in the global tale of the struggle for control of electricity grids, especially in light of the recent trends toward privatization and liberalization. This splendidly researched and easy-to-read study points out the often devastating effects of the privatization and/or liberalization of electricity markets all over the world. Curiously, however, Beder completely leaves out a major positive case of a liberalized electricity market: Germany.

By July 1, 2007, the EU will have liberalized its electricity markets. With liberalization (basically the EU term for what Americans call deregulation), state-owned power suppliers need not be privatized. Rather, to prevent cross-subsidization and to facilitate market entry for new providers, liberalization requires that the generation, transmission, and distribution of power be handled by separate organizations. Thus, large power companies cannot, for example, make competitors more expensive by selling power generation at cost and switching their profits wholly to grid services by charging high prices for transmission.

Before we move on to a chapter-by-chapter discussion of the potential of the main sources of energy and related matters such as demand management and efficiency, let us first take a general look at what Germany has done well. Germany liberalized its electricity market in 1999. The result was not massive power outages or skyrocketing electricity prices. Rather, prices fell slightly, customers were able to switch to any electricity provider they chose, the market for small producers of renewable power boomed, and the quality of the power supply remained extremely high, with outages estimated at *11 minutes per year.*

Granted, less than 3 percent of Germans have actually switched power companies, but consumers have been more actively involved in generating power themselves. Major utilities are collectively losing 1 percent of the generation market per year to small, distributed producers. Other EU countries posted higher rates of switching to renewables, such as the Netherlands, where a full 30 percent of consumers switched to green power when it was offered at the same price as conventional electricity. No wonder Robert Werner, chairman of the German renewable energy provider Greenpeace energy, has stated: "We are flying the flag of liberalization high."

Germany did not have a regulatory authority for the first six years of liberalization. In other words, liberalization has run smoothly in Germany (as it has in several other countries of continental Europe but not in the UK). While the prevalent entrepreneurial spirit in the US does not shy from creating artificial electricity shortfalls for personal profit, no one in the German electricity industry has ever tried to "take out Munich." Admittedly, at the beginning of 2004 the wind industry and consumer protectionists began complaining that the still-oligarchic electricity market — the three largest energy providers in Germany still make up around 70 percent of the market — was raising prices for no good reason. RWE, E.ON (the largest energy provider worldwide in 2004), and EnBW refuted charges of price gouging, countering that the increases were due to the growing share of renewables in the electricity supply. Once it became clear that the price hikes exceeded the extra costs of renewables by a factor of five, the inevitable was put on the agenda: Germany needed a regulatory body for its power grid.

In the summer of 2005, almost a year and a half later, RegTP, formerly the regulatory authority for telecommunications and postal services, finally added the grid to its mandate. Many Germans criticized the government and the industry for taking so long to settle the matter, but the fact is that very little was going wrong — at least from an American perspective. Germans engage in civil discussions that produce a consensus (on their news talk shows, people don't even interrupt each other, much less shout), and they had simply taken their time to do a good job when creating a grid regulator.

What about the price hikes from renewables? They are the result of the Renewable Energy Act (REA), though this law, passed by the governing coalition of the Greens and the left-of-center Social Democrats (SPD) in 1999, was essentially only a continuation and expansion of the Energy Feed-in Act that the libertarian FDP party and Helmut Kohl's conservative CDU party had

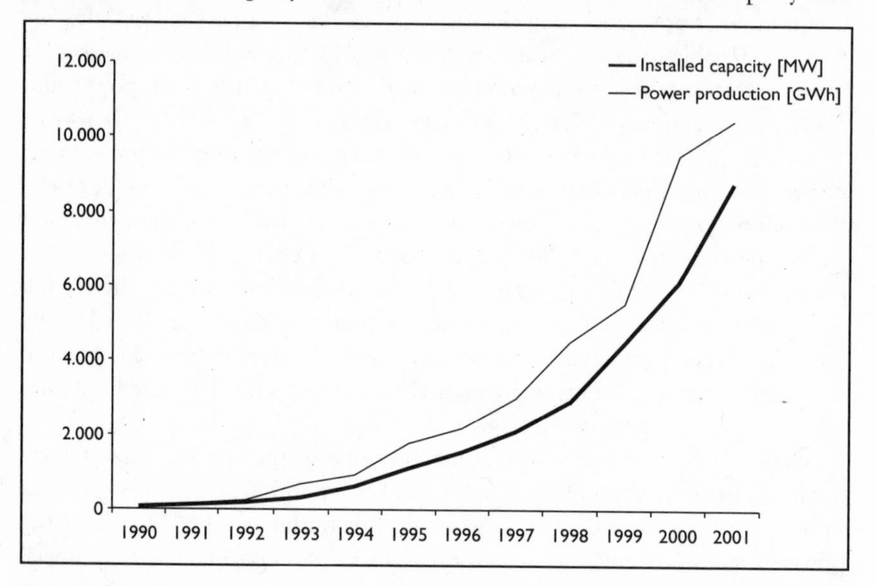

Figure 1.1:
Wind power in Germany. At the beginning of the 1990s, Helmut Kohl's government launched its support for wind energy, following Denmark's lead. In 1998, when Kohl's party was voted out of office, the success of wind energy was already clear, and the further support of the Greens/SPD in 1999 only continued the success story that the political opposition had begun. (Source: German Renewable Energy Act)

passed at the beginning of the 1990s for wind energy Here we see the major difference between the political climates in Germany and the US. In Germany there is broad support for renewables all along the political spectrum while in the US the wind and solar industries complain about the rollercoaster ride they have been on since President Carter gave them their first boost more than 25 years ago. In October 2005, the governing coalition of the Greens and the SPD was replaced by a grand coalition of the SPD and the CDU. At the time, this new coalition did not seem to want to make any immediate changes to the REA. There were only some announcements that the legislation would be reviewed after the results for 2006 were made available, but a review had been planned anyway for 2007.

It seems that the political conservatives in Germany have not estranged themselves from the concept of conservation. Indeed, Klaus Töpfer, the current and internationally respected director of the United Nations Environmental Program, is a member of the CDU. There should be no contradiction between "conservative" and "conservationist."

The REA is the key to Germany's success. Its "feed-in" system differs from the quotas and certificates used in many other countries. A price is determined that will make each type of renewable energy competitive. Retail rates are not taken into consideration. Power companies are then obligated to pay these prices to all producers of renewable energy, be they ever so small, for any electricity they sell back to the grid. Small, distributed energy producers are always paid at the same rate, whether or not they produce more energy than they consume. Anyone who wants to put a few solar panels on the roof is eligible. Indeed, small energy producers are paid the most per kilowatt-hour, with rates decreasing slightly as systems become larger up to the scale of industrial backup systems. The feed-in system made Germany the world leader in solar and wind power and is now being widely adopted in Europe.

Before we talk about the feed-in system in greater detail, let us begin by briefly describing the renewable energy credit (REC) systems common in the US today. To begin with, not every state has such a system to promote renewables, and the ones that do exist differ from state to state. Why is energy policy dealt with as a states' rights issue in the US? If the US had a long-term, nationwide renewables policy, investors could plan better. Now, people who want to invest in photovoltaics or wind energy must not only study the weather conditions around the country but also compare various programs, some of them even at the city level.

In general, RECs and renewables portfolios aim to entice industry to use renewable energy by offering tax credits and sometimes guaranteed compensation for power fed to the grid above wholesale rates — that is, above the rates paid to large, conventional power plants. A specific goal is often stated, such as a particular percentage of energy from renewables by a certain year. Texas, for example, aims to have 2,000 megawatts, or 3 percent of its capacity, from renewables by 2009. (Germany got around 11 percent of its electricity from renewables in 2005, up from 9.3 percent in 2004 and 8 percent in 2003.)

The RECs that are popular in the US sometimes play the different types of renewables off against each other, with wind energy, the one that is least expensive and therefore needs the least support, winning. This is done in the name of competition, though when we think of competition we usually think of companies rather than industries competing with each other. Boeing is not seen as a direct competitor of Amtrak or General Motors, and we should not see biomass and photovoltaics as competitors either. Competition is a given as long as the biomass and photovoltaics sectors are not monopolized. At the end of the day, what matters in many of these REC policies is that the target figure is met. If nearly all of that figure comes from wind turbines because wind power is currently the cheapest source of renewable energy in most applications, the REC policy is still considered a success, even though it fails to jump-start the very energy sources that need help the most. So if your goal is to have a broad blend of renewables, you might not be happy with RECs.

Tax incentives also are more useful if you have a high tax rate. If you are a middle-income wage earner and you decide for reasons of personal conviction to invest in renewables rather than buy a fancy car this year — an increasingly common phenomenon in Germany — you will not be receiving the same support for your investment as will a highly profitable company that could use some write-offs. The technology is the same but the support is greater when your income is higher. As a result, RECs may not lead to a widespread grassroots adoption of renewables but instead leave investments in renewables to the powerful energy companies already in charge.

Another type of support, for photovoltaics in particular, is called "net metering." Again, the details of this scheme differ from state to state and not every state has such a system. At first glance, net metering seems impressive; small power producers, such as single-family homes with photovoltaics on the roof, can sell power back to the grid by running the meter backward. This approach has many proponents, such as Jeanne Fox, president of the New Jersey Board of Public Utilities, who wrote in 2005:

> Without net metering, federal law under the Public Utility Regulatory Policy Act (PURPA) requires distributed generators to use a double meter — which separately tabulates energy consumed at the retail rate and energy produced at "avoided cost".... Double metering is also flawed because of the added expense to the utility of reading the second meter and processing a monthly check.[4]

Here Fox is opposing the practice of one of the world's leaders in renewables. In the German feed-in system, a house with photovoltaics has one meter to measure how much power is consumed from the grid and a second meter to measure how much power the solar panels produce. What difference does that make?

First, compensation for the solar power a household produces need not be linked to the retail rate of electricity in that area. In the German system, compensation is based on a rate that will make investments in photovoltaics pay for themselves in cloudy Germany over a reasonable period (not the avoided-cost rate as Fox states). Fox also seems to suggest that single metering somehow prevents utilities from lowering compensation to the "avoided-cost" level. Unfortunately, this is not true: if you produce more than you consume, your meter reads "minus X kWh" and the utility can simply say it will pay the wholesale rate for all negative figures (excess production). This is exactly what is done in some US states.

In the end, what matters is not how you meter but how much you pay for photovoltaics. What is important is not to have a target but to get all types of

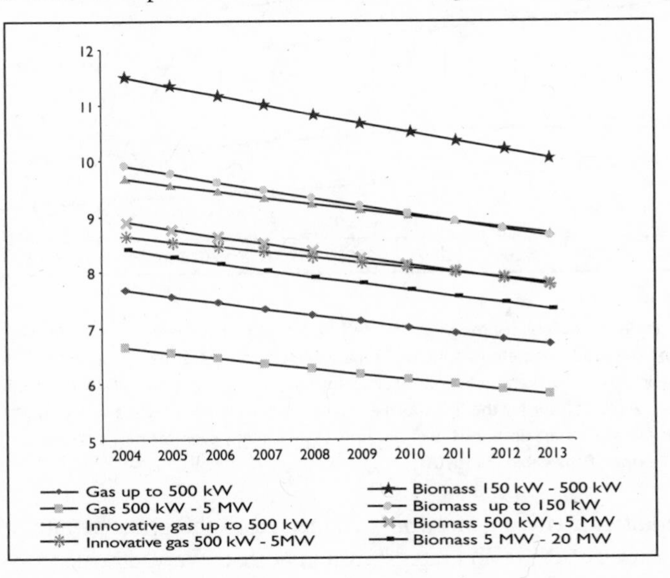

Figure 1.2:
Compensation for a kilowatt-hour of energy fed to the grid from a biomass system. A small biomass generator < 150 kW that went online in 2004 will receive around 11.5 cents for each kilowatt-hour it feeds to the grid in Germany for a guaranteed 20 years. The same system that goes online in 2013 will only receive around 10 cents per kilowatt-hour for the full 20 years. "Gas" systems here are only landfill, sewage, and mine gas, not natural gas. "Innovative" includes fuel cells, the Organic Rankine Cycle, cogen turbines, Stirling engines, and the Kalina Cycle. Note that large systems receive less compensation. The thinking here is that different systems need different levels of support and that all sources are to be supported appropriately to develop distributed generation. (Source: German Renewable Energy Act)

renewables on their feet as fast as possible. Double metering in Germany merely allows for calculations to be made separately so that compensation for power production can be separated from retail rates. German politicians sat down with industry and worked out the figures for each type of renewable energy.

As we see from Figures 1.2 through 1.5, each type of energy gets the support it initially needs to get going. Producers are paid enough to make their systems at least break even in the long term. At the same time, all of these industries are under considerable pressure to reduce prices, with the support for photovoltaics dropping by around 40 percent in less than ten years. Thus the German government is not simply throwing money at an industry until it becomes competitive but putting that industry under extreme pressure to become competitive as quickly as possible.

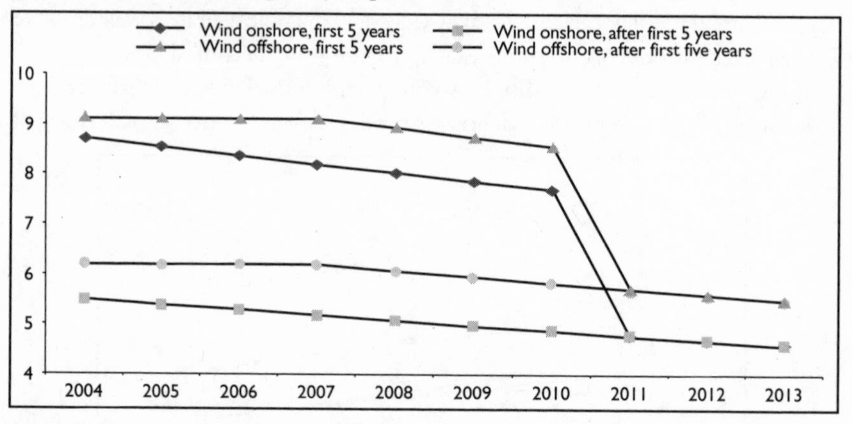

Figure 1.3:
Compensation for a kilowatt-hour of energy fed to the grid from a wind turbine. As wind turbines are already the most competitive source of renewable energy in Germany, support will drop drastically in 2011. For example, wind turbines that go online onshore in 2007 will receive around 8 cents per kilowatt-hour for the first four years but only around 5.5 cents starting in 2011, when this five-year "start-up bonus" will disappear completely, even for new systems.
(Source: German Renewable Energy Act)

In addition, once a generator is connected to the grid, the rates are stable for 20 years. The degressive rates shown indicate the level at which a system will be compensated over a period of 20 years if it goes on the grid in a particular year. In other words, a photovoltaics system that starts feeding power to the grid in 2006 at just over 40 cents per kilowatt-hour will still be receiving just over 40 cents in 2025, even though units that go online in 2013 will be receiving only just over 25 cents. This is what Germans mean by "investment security."

But there is more magic to this plan: these rates are paid for not by taxpayers but by energy consumers. The extra funds needed to compensate owners of

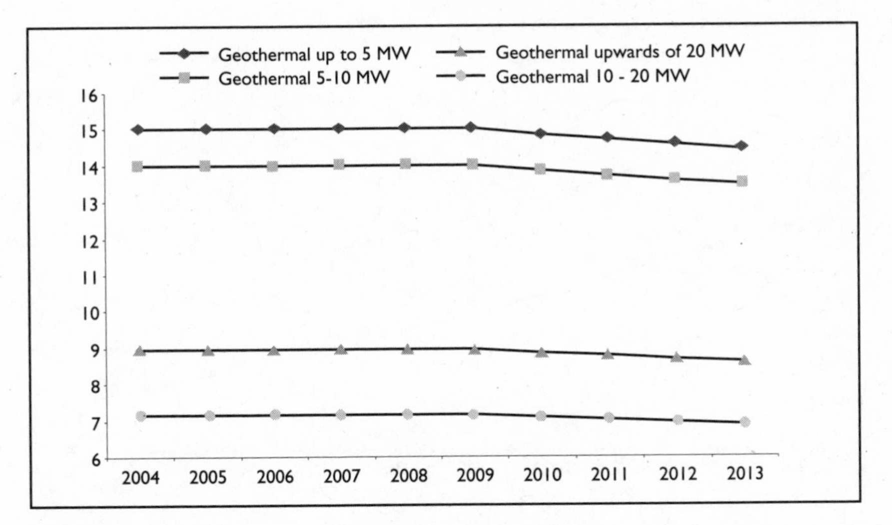

Figure 1.4:
Compensation for a kilowatt-hour of energy fed to the grid from a geothermal plant. Since geothermal plants are still in their infancy in Germany, lawmakers and industry decided to give them until 2010 to attain lower prices through technological advances.
(Source: German Renewable Energy Act)

renewable power generators are spread across all electricity consumers in a pay-as-you-go arrangement. There are two significant benefits to this approach. First, when a new political coalition enters office it has no incentive to change this scheme in order to cut governmental expenses because it is not a budget item at all. If a newly elected politician wants to make some budget cuts, there is plenty of pork in the German budget for the coal industry and the nuclear industry but there is very little to cut out of the budget for renewables aside from investments in research and development. Second, since pay-as-you-go does without money from taxpayers, the European Court of Justice ruled a few years ago that the scheme does not constitute a subsidy as defined by EU law and therefore does not distort competition.

But are these prices not outrageous? After all, power generally costs around 10 cents per kilowatt-hour in North America, approaching 20 cents only in certain areas such as New York City.* Not at all. We should not forget that power is more expensive in Europe than in the US anyway, partly because European power lines more often are buried to prevent power outages during storms. One can argue that electricity prices in North America are so low in part because such external costs as business losses from power outages and extra investments in emergency power generators are not included. Given the far higher rate of power outages in North America than in Germany, these exter-

*I have chosen to state all figures, whether euros or dollars, in cents as though there were parity between the two currencies. Indeed, considered over the years since the euro was introduced, there is near parity. Providing the figures in cents will allow the reader to use conversion rates valid at the time of reading.

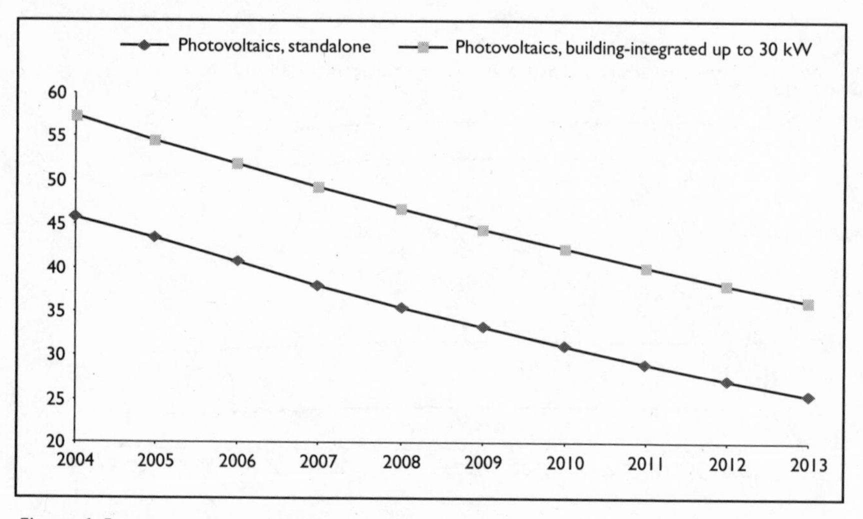

Figure 1.5:
Compensation for a kilowatt-hour of energy fed to the grid from a PV system. The rates provided for photovoltaics (PV) in Germany are among the highest in the world but they are also the most degressive, dropping by 6.5 percent per annum for stand-alone systems as of 2006. Germany calculated that PV would need some five times as much support as wind and designed rates accordingly. In addition, because façade systems are considered desirable, an extra 5 cents per kilowatt-hour is paid on top of the building-integrated rate (see Chapter 9). (Source: German Renewable Energy Act)

nal costs can be assumed to be much higher in North America. Business losses exceed the reduction in investments to improve the grid by at least a factor of 1,000, according to one technology website:

> In every year since 1975, about $117 million less was invested in the electricity transmission grid, according to the US Department of Energy. The result is a brittle grid constantly on the edge of failure, with research firm Primen estimating the annual cost of power outages and fluctuations at between $119 billion and $188 billion yearly.[5]

Some have estimated these costs a bit lower, such as Joseph Eto of the Lawrence Berkeley National Laboratory, who put the figure at $80 billion per year.[6] But whatever the exact figure, Germans have far lower losses from power outages and are not worried about making power uncompetitively expensive. According to a survey conducted by a German news channel in 2004, four out of five Germans support pay-as-you-go for renewables. Of those surveyed, 59 percent even wanted this support to be expanded, while only 12 percent thought it should be reduced and 26 percent wanted to keep support at its current level.[7] Not only do the degressive rates put companies under pressure to lower prices quickly but also, as the still-small share of renewables in the overall electricity supply (around 1 percent for Germany at the end of 2005) increases by leaps and bounds because of this degression, the burden on consumers will grow only slightly. At the end of 2005, the extra costs for renewables fed to the grid amounted to only around 1 euro per month for the average German family. Of the 19 cents per kilowatt-hour charged in Germany, around 0.5 cents stems from the feed-in of renewables (though RWE was still charging an extra 0.66 cents in 2005).

The emphasis on small units, which generally receive greater compensation, also makes sense politically. The more citizens become involved in investments in renewable energy, the greater public support will be. In Germany, in addition to rooftops of single-family homes covered with photovoltaics one often finds small groups of only a few wind turbines or even a single turbine. Notice that German law often makes distinctions between units as small as 150 and 500 kilowatts. In contrast, the REC legislation in Texas defines a small producer as one that is less than two megawatts in size.

The emphasis on tax credits in the US means that the focus will be on large wind farms developed by large utilities. One undesirable and perhaps unforeseen effect is that local residents may come to perceive such projects as merely an intervention in their local landscape rather than something they can be a part of. Facilitating the participation of small groups of citizens — all the way down to single families and individuals — will literally empower citizens and increase general support for renewables, which then will not be seen as a necessary evil that, like coal and nuclear plants, may have to be put somewhere but better not be "in my backyard." Encourage people to invest in renewables and they will put lots of them in their backyards.

In addition, the installation of small units will take a further burden off the power grid. Large wind farms produce a tremendous amount of energy, possibly in remote areas or offshore, where wind conditions may be good but power consumers are few. Such projects are a burden on the power grid, which may have to be expanded just as when a coal or nuclear plant is built. Not so with small power generators close to consumers (including small wind farms or single turbines). Such distributed power does not really need the grid, which then is mainly used for backup power.

We will go into these issues in more detail in the following chapters. In concluding this introduction, let us return to the challenge posed by Shellenberger and Nordhaus, who call on environmentalists to present a positive vision of the future that encompasses all areas of life — not just the micromanagement focal points such as gas mileage, emissions certificates, and mercury levels in bodies of water. The authors ask: What if Martin Luther King Jr. had said "I have a nightmare" instead of "I have a dream"?

What, then, is the European dream? Imagine a world with very few resources. People have come to understand and accept that the remaining finite resources will have to be devoted to setting up a sustainable, renewable system while we still can. Buildings have become largely energy-independent, with scarce fossil resources being used only as a last resort. People are not worried about higher prices because they understand that imports of fossil energy have given way to domestic jobs. Farmers grow not only food but also energy. The environment is cleaner and there is full employment because renewables are job-intensive.

This switch from finite to renewable energy sources required plenty of innovation. People understood this and invested heavily in the education of their children — not in a select group of elite schools but in public education for the masses because everyone was needed in this massive transformation. The goal was not the American dream that anyone can become rich but rather that the largest possible number of people should lead happy lives in good health, free from job-market anxiety and able to make a contribution to society.

The power grid was understood as part of the commons. Preference was given to renewable resources, with finite resources being used only when there was not enough renewable energy. Of course, powerful energy companies initially fought this development but people came to realize how easy it is to pay them off. The ultimate business goal of private utilities is not, after all, to make electricity but to make money. Electricity and oil are merely mediums.

Energy companies began complaining that small, distributed generators relegated them to the status of emergency power producers and raised costs by causing their plants to run below capacity because renewables had top priority. People responded by setting up a government agency to work out appropriate compensation for the utilities. "If utilities are losing money," the citizens said, "we'll pay them." People simply wanted to take control of their own lives, and being seen by large companies as power generators was an important step in that direction. The companies themselves soon realized that no one was stopping them from investing in renewables.

Again, few consumers complained about the resulting price hikes because most understood that prices are not the same as costs. If prices rise, you can change your behavior and technology to lower costs. People did not feel they were giving something up by not driving off-road vehicles on roads; they felt they were improving their health by walking to work in compact cities and saving money by not needing a car for everyone in the family.

Indeed, to slow down consumption people even voluntarily raised the prices of finite resources through higher taxes. They also understood that the revenue from these taxes did not have to be used for traditional "environmental" purposes, such as wildlife preservation, to make such taxation ecological. After all, wildlife already benefits from a cleaner environment. Rather, this revenue was used to develop public transport, lower social security payments to encourage businesses to hire more people, and otherwise allow the poor to accommodate the higher prices.

All of this required serious government regulation, which in turn gave birth to a new breed of politicians. People stopped electing those who promised to get the government out of their lives and started voting for those who said that government services should actively make the lives of citizens better.

Self-sufficiency, conservation, independence, peace, hard work, innovation, freedom: do these sound like American values? Then read on to see what you think Americans can learn from the progress of Germany and the EU toward the realization of their dream.

Chapter 2

The Big Picture

I feel that long-term energy security is more important than the short-term illusion of somewhat lower energy prices. Those who criticize renewables for allegedly being too expensive are just looking for a scapegoat.
—Norbert Walter, head economist at Deutsche Bank, 2005

IN THE MIDDLE AGES, people thought that the silver taken out of mines would grow back again. And when coal began to be used in northern England on a large scale in the 14th century, people would smear manure on the coal seams to make them grow faster.

We have come a long way since then. True, we do argue about whether we will have enough oil for another 10, 20, 30, or more years and whether coal will run out completely in 200 or 300 years. But there is general agreement that we are using these valuable finite resources much faster than they could ever regenerate. According to some estimates, we burn in one year what it took nature 15,000 years to make.

And yet, future generations are unlikely to find us much smarter than the people in the Middle Ages, for we are not putting our knowledge to work. Germans cannot understand why so many Americans need SUVs on paved roads, but we should not forget how long it took Germans to make catalytic converters mandatory. Our descendants probably will not see those who inhabited industrialized countries in the 20th and 21st centuries as the enlightened souls we hold ourselves to be. More likely they will see us as the most egocentric people in the history of humankind. Probably no cultures have ever endangered the standard of living of future generations as much as we have in order to make our own lives "better." Our children and our children's children will inherit from us an impoverished world — and not just in terms of energy.

A reader today of Herman Melville's famous novel *Moby Dick* (1851) might be astonished by the narrator's comment that "we account the whale immortal in his species." The narrator was convinced that, despite all threats to its existence,

> "The eternal whale will still survive, and rearing upon the top-most crest of the equatorial flood, spout his frothed defiance to the skies."[1]

Today we know that whales are not safe anywhere from pursuit by humans. No more than two million sperm whales, the largest carnivores alive today, still exist and the species is considered endangered. Though commercial whale fishing has been banned for some 20 years, the numbers of most species that have survived are so small that populations are having a hard time recovering. And the situation is not better for other sea animals: a recent study found that about 90 percent of all large fish (such as tuna and swordfish) have been wiped out by industrial fishing.[2]

Indeed, the Earth's resources are vanishing in almost all areas — from metals (most of which should last for another five to seven centuries) to biodiversity. We are currently experiencing the largest mass extinction of species in the past 65 million years and the main cause is civilization. The fourth *Millennium Ecosystem Assessment*, published in 2005, found that some 60 percent of the ecosystems studied were being degraded.[3]

This topic goes beyond the scope of this book, but it is appropriate that we begin here so that we do not forget the big picture. Our energy problems are part of our larger resource problems, which represent a moral dilemma that almost no one is addressing. Developing nations may not be able to assert their interests as well as industrialized nations, but the people in developing nations are by no means voiceless. They may put themselves in harm's way when they take to the streets to protest, as recently seen in the popular opposition to Bolivia's energy exports which were nationalized in May 2006, but at least they can do so. Who is going to stand up and speak on behalf of the people who will be alive in 2200?

Today we have one answer (for which French critics have coined the term *pensée unique*) to all questions about the future: the market will take care of it. This neoliberal approach, also known as TINA for "There Is No Alternative," makes the resource problem into a problem of cost, which is in turn a moral issue. Any discussion about money is about distribution, especially when people call for growth instead of redistribution.

On the other hand, in recent years more and more environmentalists have been discovering that the market can be used for their purposes too. In fact, anyone who wants to make the world a better place had better take account of the "free"-market economy. Some environmentalists are still at odds with market mechanisms but there is no need to be. Indeed, it is much easier today to use supply and demand to promote sustainability than to toss out free-market enterprise and try to set up another system entirely.

Tax the Bads, Not the Goods

Ecological tax reform aims to make people's consumption and behavior more environmentally friendly by making products and services with a great environmental impact more expensive. In other words, we tax bad things, not good things. Germany has been a leader in this field. Its *Ökosteuer* (eco-tax) took effect in 1999 to make various types of non-renewable energy more expensive. This additional tax revenue was then used to lower non-wage costs (such as employer contributions to social security and health insurance) that make German workers expensive. For example, for five successive years roughly one cent was added to each liter of fuel, effectively raising the price of a liter of fuel by almost 5 cents over five years equivalent to an increase of 18 cents/gallon. This gradual approach sent a signal to consumers, who were able to plan ahead in anticipation of rising fuel costs without suffering from any sudden price shocks.

Many Germans wondered why the revenue from an eco-tax would be used for purposes that have so little to do with the environment. However, ecological tax reform is generally designed to be revenue-neutral. In other words, neither the government's budget nor the burden on taxpayers should be increased. Rather, the current level of taxation is restructured so that more taxes come from activities that society considers environmentally undesirable.

Before we discuss whether the German eco-tax produced the desired results, let us first take a look at some options chosen elsewhere and their results. Assuming that we want to raise the gas mileage of new cars sold, we could try to pass regulations forcing the automotive industry to manufacture such cars. The United States has been pursuing this approach with its Corporate Average Fuel Economy (CAFE) standards,[4] and the results have been disappointing. Gas mileage in the US has risen slightly since the mid-1980s but people can afford lots of luxury extras such as air conditioners and powerful engines, which have more than offset any increases in technological efficiency.

California tried a similar approach when it ruled that 10 percent of cars sold in the state would have to be emission-free by 2003. In the end, industry was able to sit back and say, "We can produce battery-powered cars but consumers won't buy them." (It also was hoped that fuel cell cars would be available by the beginning of the 21st century, but the automotive industry missed this goal completely, as we shall see in Chapter 11.) This focus on a radically different technology that did without fossil fuels completely also shifted the focus away from solutions such as hybrid engines, with which Japanese manufacturers have been very successful.

Environmental tax reform comes at the problem from another angle entirely. To begin with, let us make one crucial distinction: prices must not be confused with costs. For example, as many North Americans are aware, gasoline prices are roughly twice as high in the EU as in North America. But North

Americans spend about the same on gasoline as Europeans do, mostly because of urban sprawl and cars with much lower gas mileage. So while gas *prices* in North America may be half those in the EU, North Americans have the same *costs* because they consume twice as much. North Americans can thus afford to waste energy, which is why the Global Governance Project found in its *Implementing the Kyoto Protocol Without the United States: The Strategic Role of Energy Tax Adjustments at the Border*[5] of 2003 that low energy prices in Canada and the US constituted unfair competition.

Supporters of the eco-tax also believe that the market will take care of everything, but — unlike many neoliberal economists — they do not see the market as some natural force that automatically takes over when the government stops intervening. On the contrary, proponents of the eco-tax use the market as an especially effective instrument to change people's behavior.

Was the German Eco-Tax a Success?

For the first time in the history of the Federal Republic of Germany, fuel consumption fell in four consecutive years (2000-2003) after the start of the eco-tax. Sales of thriftier cars rose, as did the number of people using public transport, the latter rising every year for five consecutive years (1999-2003).[6] According to a poll published in the German magazine *stern*, 87 percent of those surveyed wanted their next car to have better gas mileage.[7]

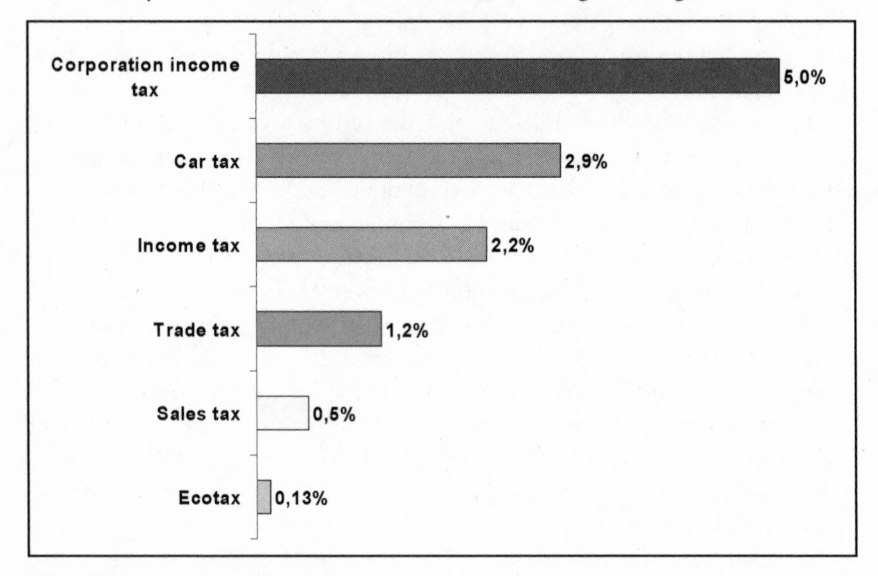

Figure 2.1:
The administrative costs of various types of taxes as a percentage of revenue in Germany. The eco-tax performs the best here. Almost all of the revenue from the eco-tax is used for its intended purpose. (Source: Green Budget Germany)

Here we begin to understand better why the eco-tax makes environmental sense even though the revenue is used not for environmental projects but to lower non-wage costs. Environmental protection is important, but so is Germany's unemployment rate (above 10 percent for many years now). Making German labor less expensive by lowering non-wage costs should help create jobs. Also, higher energy prices will reduce fossil energy consumption, a result that is environmentally friendly regardless of what is done with the tax revenue.

Of course, revenue from environmental taxation can be used for environmental purposes. In practice, however, higher energy prices may be a greater burden on the poor, who are less able to buy an expensive new hybrid car to compensate for rising gas prices or to insulate their homes, which they may not even own. Using the eco-tax to create jobs will help reduce this burden.

In North America, where public transport is often poorly planned if not lacking entirely, it might be best to invest the revenue from an eco-tax in developing and expanding public transportation services. Such an approach would not only provide an affordable alternative for those not able to invest large sums in energy efficiency but also reverse the dismantling of public transportation of the mid-20th century, when rubber companies and the automotive industry bought up streetcar lines in order to tear out the tracks and pave the way for buses and cars.

At the same time, we should keep in mind that the goal is not to have energy prices continue to rise without end but rather to reach a switchover point where using efficient, environmentally friendly technologies no longer makes any significant financial difference. Gasoline and diesel should not be cheaper than biofuels, flying not cheaper than taking the train. The Netherlands is a good example of what happens when a switchover point has been reached. There, green electricity has not been more expensive than conventional electricity for several years, and the Dutch cannot save money by not buying green power. As a result, 30 percent have chosen green power. Demand now exceeds domestic supply and green power has to be imported, mostly from Germany.

Growth

Ecological tax reform combines environmental awareness with a belief in the much touted invisible hand of the market. There thus is no inherent contradiction in support for both market forces and environmental conservation. In fact, Green Budget Germany, the organization that helped design Germany's eco-tax, awards a prize for environmental economics every year: the Adam Smith Prize, in honor of the Scotsman who coined the term "invisible hand."

In the US, many environmentalists also have no trouble embracing market forces. One prominent example is Robert Kennedy Jr., one of the best-known environmental protectionists in the US and a major critic of the environmental policies of the Bush administration. As he put it in an interview with *Grist Magazine:*

> The best thing that could happen to the environment is free-market capitalism. In a true free-market economy, you can't make yourself rich without making your neighbors rich and without enriching your community. In a true free-market economy, you get efficiencies and efficiency means the elimination of waste. Waste is pollution. So in true free-market capitalism, you eliminate pollution and you properly value our natural resources so you won't cut them down. What polluters do is escape the discipline of the free market. You show me a polluter, I'll show you a subsidy — a fat cat who's using political clout to escape the discipline of the free market.[8]

The main difference between proponents of eco-taxation and the neoliberal economists normally associated with the philosophy of Adam Smith is the neoliberal view of the invisible hand as a sort of magical force that we cannot, by definition, control. According to the most common misreading of Adam Smith's work, as soon as we try to regulate the market the invisible hand stops working. There is a lot to criticize about this interpretation of Adam Smith's work, starting with the simple fact that Smith himself never intended the term "invisible hand" to be used as neoliberals use it today.[9]

Second, a market on which the invisible hand could work would require that all players be completely, or at least equally, informed. In reality, to take one pertinent example, few people realize that an energy-efficient lamp may cost a bit more at the store but also may quickly pay for itself by lowering their energy bill. In many cases, when we purchase an appliance we do not think about how much power it will consume and what that will cost us. But when it comes time to build a new nuclear or coal plant, everyone seems to be against it.

Third, growth is not necessarily good. If I can't get a job near my home, I may have to buy a car so I can get to work. I may also not have much time to eat breakfast, so maybe I'll drop by a fast-food chain and pick up something to eat on the road. And while I am eating with one hand and talking on my cell phone with the other, I may have an accident, and I will have to repair my car. This is not my idea of a happy life, but each of these events promotes economic growth — at least as it is measured today. If I make my breakfast at home, I spend a lot less money; if I can walk to work, I may be able to do without a car altogether, and even less money will be circulating. But I will be happier and my health probably will be better. In other words, what is good for economic growth is not necessarily good for me. Economic growth therefore cannot be the answer to all our questions.

Fourth, there are costs that the market does not take into consideration when prices are set. These are commonly referred to as "external costs." For example, environmental impacts are often not included when energy prices are calculated. We thus often speak of a kilowatt-hour of coal power costing only around 3 or 4 cents, but this low price is possible only because, for instance, we do not force utility companies to restore the land devastated by mountaintop removal (see Chapter 5) and strip mining.

The concept of external costs is closely related to environmental protection. At the beginning of the 20th century, British economist Arthur Cecil Pigou pointed out that emissions from coal plants led to costs for third parties. Back then, the cities of newly industrialized England were under a cloud of smog. The "Pigou tax" was scoffed at as a curiosity, but it laid the foundation for what we now know as the polluter-pays principle.

Energy carrier	External costs of electricity generation in cents per kWh
Coal	4-8
Oil	3-5
Natural gas	2-3
Biomass	<1-3
Nuclear	<1
Hydro	<1
PV	<1
Wind	<1

Figure 2.2:
The external costs of electricity generation in cents per kilowatt-hour. The figures are rounded off and based on the 2003 study by EU DG Research *External Costs: Research Results on Socio-Environmental Damages Due to Electricity and Transport*. In contrast, Ari Rabl and Joseph V. Spadaro ("Public Health Impact of Air Pollution and Implications for the Energy System." *Annual Review of Energy and the Environment*. Vol. 25, Nov 2000, pp. 601-627) estimate the external costs for coal power at 4.54 euros (not cents!) per kilowatt-hour and for natural gas at 1.12 euros per kilowatt-hour. Germany's Wuppertal Institute came to 26 cents per kilowatt-hour in external costs from brown coal in its October 2004 study Brown Coal: *A Source of Energy*

These are only a few of the weak points in neoliberal ideology, and I have only gone into them cursorily. Perhaps the most important point to remember is that neoliberalism does not know any limits to growth. Take, for example, the question of oil depletion, which we will come back to in Chapter 3. Neoliberal economists say we don't have to worry: as oil prices rise, exploration will pay for itself more and oil fields and unconventional energy sources such as oil shale will become competitive. But rising prices will not put more oil in the ground. At some point — and there is a great debate over when this will happen — current production will not be able to keep up with current demand. This point is commonly referred to as "peak oil." If there is a rush during that peak to bring to market everything that can possibly be refined, we may very quickly deplete what is left. The peak itself would then be extended, but the downward slope after the peak would be all the more precipitous.

Some economists, most notably Herman Daly (formerly of the World Bank),[10] try to point out that growth cannot go on forever since the Earth obviously ends somewhere, but such arguments are belittled in neoliberal circles. This irrefutable fact erodes the foundation of their economic system of beliefs. Neoliberal counter that it may be true, that the earth has limited resources, but that does not tell us very much. And they are right: what we need to know is whether we are living beyond our means today or whether we can keep on growing globally at around 5 percent (2004) for the next 100,000 years.

The authors of *Limits to Growth*[11] have been trying since 1972 to find out what the outcome might look like. Unfortunately, their works have been widely misread, in some cases perhaps intentionally. The most common misreading is that the authors claimed in 1972 that the world would come to an end in 30 years: too many people, no more oil. This obviously didn't happen, and through the years of prosperity since then *Limits to Growth* has been taken as the best example of how far environmental protectionists are from reality. They allegedly are against the very concept of progress and would just as soon have us return to the lifestyle of the Stone Age. Recently, Danish economist Björn Lomborg argued in his *Skeptical Environmentalist*[12] that the environment is actually getting better.

Unfortunately, proponents of the theory of unlimited growth seem to have unrealistic expectations of what technology can do. One prominent example is Julian Simon, who entered into a bet with Paul Ehrlich after the second oil crisis of 1979. Simon bet that the prices of five types of metal, which he would even allow Ehrlich to choose, would drop between 1980 and 1990, while Ehrlich held that scarcity would drive prices up. Simon reasoned that technology would solve all our problems:

> Technology exists now to produce in virtually inexhaustible quantities just about all the products made by nature — foodstuffs, oil, even pearls and diamonds — and make them cheaper in most cases than the cost of gathering them in their natural state.[13]

Ehrlich was known for forecasting a major catastrophe in the late 1960s caused by the explosion in population growth, which he predicted would

ECOLOGICAL FOOTPRINT

In their 1996 book *Our Ecological Footprint: Reducing Human Impact on the Earth,*[a] Mathis Wackernagel and William Rees coined the term "ecological footprint" to describe the area that a group of people need to maintain their standard of living. As we can see from Figure 2.3, which shows us the ecological footprint of Santa Monica, if Santa Monica had to take all its resources from the surrounding area no one would be able to live in Los Angeles, Santa Clarita, Malibu, or Long Beach.

At the moment, the West depends on imported resources. Worldwide, Wackernagel and Rees estimate that we currently are overshooting the carrying capacity of the Earth by 35 percent. But figures vary: according to the World Wildlife Fund's *Living Planet Report* of 2004,[b] the Earth's carrying capacity is exceeded by "only" about 20 percent. It is estimated that roughly 1.8 hectares are available for every human now alive, but our current lifestyle requires 2.2 hectares. According to the WWF, an American requires 9.5 hectares, twice as much as a western European.

If we listened to those in developing countries more, we might have been able to recognize sooner that we will not all be able to lead the lives of colonizers. As Mahatma Gandhi once said, "God forbid that India should ever take to industrialism after the manner of the West .... It took Britain half the resources of the planet to achieve this prosperity. How many planets will a country like India require?"[c]

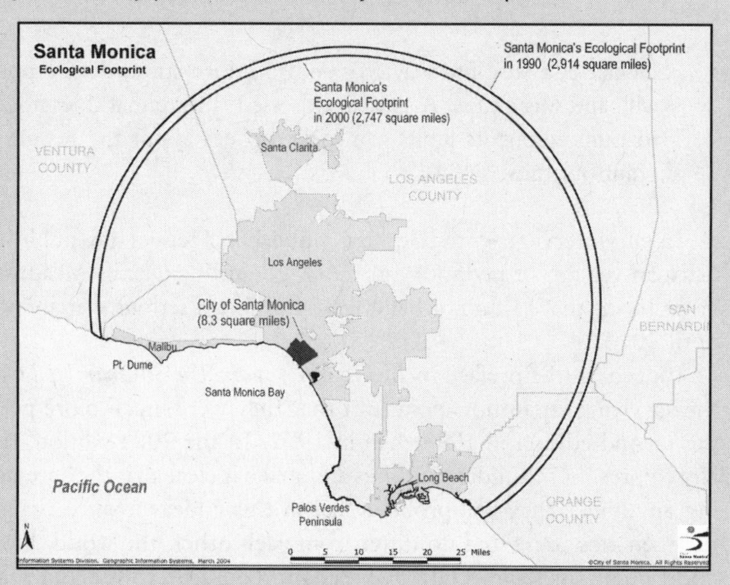

Figure 2.3. *Ecological Footprint of Santa Monica, California (Source: City of Santa Monica)*

a. Wackernagel, Mathis and William Rees. *Our Ecological Footprint: Reducing Human Impact on the Earth.* New Society Publishers, 1996

b. See http://assets.panda.org/downloads/lpr2004.pdf, cited Feb 8, 2006

c. See www.economist.com/surveys/displayStory.cfm?story_id=1199867, cited Feb 8, 2006

sooner or later outstrip food production. This has not yet happened, and Ehrlich also lost the bet with Simon for all five metals. Doesn't that just go to show that those who claim the sky is falling because of climate change, overpopulation, the scarcity of resources, or whatever the *catastrophe du jour* is are just modern-day descendants of the people who ran around in the Middle Ages telling everyone that the end of the world was nigh?

Perhaps not, for the *Limits to Growth* authors point out that they have been misunderstood. Nowhere did they predict anything, much less the end of the world in 30 years and not even the end of oil in 30 years. In fact, in stating that oil "reserves" would run out in 30 years, they were only citing figures from major oil companies. (For a definition and discussion of oil reserves, see Chapter 3.) They did not try predict the future but merely changed the parameters in their computer program to see what the world might look like, say, if ten billion people were living on the Earth by 2020, many of them in the fast-growing countries in Asia instead of in Africa. In other words, there were many scenarios. One of the authors, the late Donella Meadows, defended herself and her colleagues a few years ago, showing again how unfair it is to blame US environmentalists for the shortcomings of US policy:

> Denial is a sure-fire way to confuse information, defuse political will, and waste time. A growth-obsessed culture that does not want to think about its limits can make up lies about the people who point out those limits.[14]

In an interview with German publisher Heise (the publisher of the German version of my book) in 1998, co-author Dennis Meadows went so far as to say that "I do not believe we will face a serious scarcity of resources until 2012."[15]

Indeed, in the preface to *Limits of Growth: The 30-Year Update* of 2004[16] the surviving co-authors point out that they were much more pessimistic in the second edition in 1992 than in 1972. In the 2004 edition, they mainly aim to prevent misunderstandings and make it clear that they are not predicting anything. They will probably fail in this endeavor once again for while the scenarios presented do differ from each other, the world does not look good in any of them. Only one scenario could be considered slightly optimistic. In a second scenario, the world would be acceptable, though poorer. In the other seven, the question is only whether the planet will collapse before or after 2050. There is an "overshoot" in all nine scenarios. But before we talk about what that means, let us return briefly to the alternatives to growth.

Development Versus Growth

Critics of neoliberalism do not have anything against growth. Growth can result in greater prosperity but more often than not it seems to increase the gap between rich and poor. As a result, many economists are shifting their emphasis from growth to development. Development means that standards of living improve. Given greater efficiency, such improvements certainly are possible without growth. Indeed, if efficiency increases fast enough, standards of living can improve even if the economy is shrinking. French professor Serge Latouche has coined the term *décroissance* (degrowth).[17]

In the period after World War II, economist Nicholas Georgescu-Roegen defined the purpose of economic activity with a surprisingly simple equation that serves as the basis for much of the current criticism of neoliberalism.[18] On one end, raw materials and resources are input; on the other end, waste is output. The goal of this process is "happiness" in the middle. We therefore should see to it that we get as much enjoyment out of the fewest resources and the least waste possible. Here, economic activity is a destructive process, for in the end all products are waste.

Let us not forget that the reduction in the planet's resources is not included in any calculation of gross domestic product (GDP). We are depleting resources and calling it growth. This criticism has led to the development of alternative indicators to gross domestic product, such as the genuine progress indicator (GPI), which subtracts the detrimental effects of economic growth (such as pollution) from GDP. Indeed, the basic problem with measuring GDP is that nothing is ever subtracted. There is no count of a country's resources, so there cannot be a record of the percentage of resources depleted in any given year. It could very well be that resource depletion exceeds growth in domestic product. Who knows?

This problem is very well illustrated by the annual *Vital Signs* published by Worldwatch Institute.[19] While the 2005 edition contains charts showing everything from meat production to wind energy generating capacity and, of course, both GDP and GPI, there is no chart indicating how many fish are in the sea, how many are reproduced naturally every year, and how many we catch. Rather, there is a chart showing the world fish harvest in terms of wild catch and aquaculture. In other words, there is addition but no subtraction.

I am not criticizing Worldwatch Institute here. In all their statistics they are merely quoting the figures collected by other institutes and organizations, and they do that very well while also providing insightful analyses. Rather, the shortcomings of their *Vital Signs* are indicative of a greater problem: since we are not keeping track of how much the Earth "reproduces" each year, we cannot tell for certain whether we are taking too much. Indeed, the discrepancy in the extent to which we are overburdening the planet's carrying capacity — 35 percent or 20 percent? — is so great that the unconverted must be unconvinced.

What we need, then, is something more along the lines of what Worldwatch includes under "Environment Features" — for instance, the chapter "Forest Loss Continues." Here we read that "global forest cover stands at approximately half the original extent of 8000 years ago,"[20] which is at least the kind of subtraction we are looking for, though the time frame is so large that it does not tell us whether most of the damage occurred during industrialization or perhaps is due simply to changes that have gradually taken place since the last Ice Age. Fortunately, Worldwatch also cites figures from the UN Food and Agriculture Organization stating that the forest area lost globally every year is roughly the size of Portugal (though the statistics are from 2000). But there is no reliable global estimate of extent of forest depletion, as we will see in our discussion of biomass in Chapter 4.

To return to the debate between Simon and Ehrlich: while it is true that prices for oil and metals fell between 1980 and 2000 (adjusted for inflation), there can also be no doubt that we have less oil and metal than 25 years ago. After all, prices reflect not how much there still is but only supply and demand. Prices will not skyrocket until supply cannot keep up with demand.

That may happen much closer to the end than many people expect. The authors of *Limits to Growth* use the following example: if a plant in a pond takes up twice as much space every day and will take 30 days to cover the pond completely, when will you notice the plant? On the 29th day, the plant will cover only half the pond; on the 28th day, only a fourth. You probably wouldn't even notice the plant until a few days before it is too late to react.

It is a nice anecdote, but the facts tend to be a bit messier. Take, for example, the Sixth Extinction, as the largest wave of extinctions since the dinosaurs that is now going on is being called. *Limits to Growth* points out that we don't even have a rough idea of how many species there are on the planet. Estimates range from three million to 30 million — which is why the authors are quick to point out that "we cannot know these numbers for sure."

OVERSHOOT, COLLAPSE, AND ENERGY BALANCE

Let us now come back to the number of fish in the sea to explain one of the most important terms in this book: energy balance. For the past five years, the global fish catch has been stagnating. Some fear it may start to drop. Once we start catching more fish than can naturally reproduce, the basis begins to shrink or, as the authors of *Limits to Growth* put it, "erodes." At that point, we begin to live not only from the interest but also from the capital itself.

Nevertheless, we have no trouble buying enough fish on the market, although it is not getting any cheaper. Part of the reason is that aquaculture has been booming in recent years. Unfortunately, aquaculture is not the answer to our problems but will only increase them, perhaps even drastically, for as *Limits to Growth* states, wild fish are a food source, while cultivated fish are a food sink. Wild fish eat plants and animals that people generally do not, but farmed fish consume more high-quality food than they provide. In other words, the energy balance is negative for aquaculture. Similarly, the energy balance is negative for non-renewable energy sources.

The question is whether we will realize when we have overshot our limits. As we will see when we come back to the issue of peak oil, there is a window of time in which we can react. As *Limits to Growth* says: "Oil depletion will not appear as a complete stop." At the moment, there seems to be sufficient reason to believe that we are living beyond our means. But regardless of where one comes down on this issue, one thing is clear: we have to keep a close eye on our resources so we can react quickly to prevent them from collapsing. If we don't, the world will be a much poorer place only one or two generations from now.

The Energy Mix and the Transition from Fossil Energy to Renewables

But let's get back to energy. Oil is already becoming expensive and most experts believe that demand will outstrip supply once and for all within the next 10 to 20 years. Some even think this already has started to happen. After that, global supplies of natural gas will peak; regionally, supplies are already short, such as in the United States, which put all its eggs in the natural gas basket in the 1990s without thinking about where it was going to be getting the fuel in the future. Right now, the US is very dependent on supplies of natural gas from Canada and is also promoting liquefied natural gas (LNG) to make shipments of natural gas from Southeast Asia economical.

The focus in the US is now turning toward more coal plants and possibly nuclear power. The US has been called the "Saudi Arabia of coal," with supplies that probably will last for several hundred years at current rates of consumption. Critics of both of these energy sources believe they will spell the end of the world as we know it: coal because of global warming and nuclear power because of meltdowns and atomic waste. We will come back to the potential of coal, its drawbacks and its benefits, in Chapter 5 and will deal with nuclear power in Chapter 6. For now, suffice it to say that the proponents of each of these options argue that renewables are not a realistic alternative.

We will be taking a look in separate chapters at the main types of renewable energy. Critics of renewables point out that they are expensive and that most of the technologies are not yet fully developed. We will therefore evaluate how mature these technologies are and whether the high costs are reasonable. Critics also note that some sources of renewable energy cannot be controlled.

We have solar power when the sun is shining and wind power when the wind is blowing but not necessarily when we need power.

It must be stated at the outset that this book does not propose a single source of energy as the solution. Rather, the solution will be found in a mixture of energy sources and the particular mixture will differ from country to country. The weakness of one source of renewable energy may be easily compensated for by others. For example, wind and solar power can complement each other: when clouds gather overhead, blocking out the sun, the wind usually stirs up.

Attentive readers will have noticed that we have been talking mostly about electricity. What will we do to replace oil, the main motive fuel and the first source of energy that will become scarce? Indeed, we are facing quite a dilemma, for it seems easier to generate electricity and heat than to find a fuel that is as easy to transport and has such a high energy density as oil. Two candidates are hydrogen and biomass (in North America, mostly ethanol; in Europe, mostly biodiesel). Each of these options will be discussed in a separate chapter. But I don't want to leave you hanging: there is no easy solution, which is especially unfortunate as oil will be the first problem the world will have to address.

Renewables have one major advantage: we cannot overuse them (with the possible exception of geothermal and biomass, as discussed in their respective chapters). We cannot exhaust the sun by using too much solar power and the amount of wind we could take out of the atmosphere with wind turbines is negligible. The same is true for tidal power plants, which will hardly be able to shake the moon out of its orbit.

In contrast, we are running out of oil, gas, coal, and uranium. According to a study by MIT,[21] the world's uranium mines have not been able to keep up with demand for many years, which is why the amount of uranium in stock has dropped by around 50 percent from the level of 1985. According to the World Nuclear Association, demand will exceed supply by another 11 percent by 2013.[22] No one is talking about it, but nuclear power is overshooting and the uranium feedstock may collapse if we continue to build nuclear plants.

Before we move on to discuss individual sources of energy and the related issues of efficiency and demand management, let us define a goal. People are coming to realize that the Fossil Age will be short, probably lasting only a few centuries. We likely are in the middle of this age and we have to use the time and fossil resources that remain to set up a sustainable, renewable system. As Dennis Meadows stated in an interview with the German weekly *Die Zeit,* "Nonrenewable resources, such as the oil reserves in the Earth's crust, must not be consumed faster than we develop renewable alternatives like solar energy."[23]

Some experts, such as Donald Aitken, author of the White Paper written for the International Solar Energy Society, believe we are running out of time:

> The window of time during which convenient and affordable fossil energy resources are available to build the new technologies and devices to power a sustained and orderly energy transition is short ....[24]

Or, as Hermann Scheer, a German politician and energy expert who won the Alternative Nobel Prize in 1999, writes in his *Solar Manifesto:*

> Without a radical shift of the world's energy supply system to non-destructive solar energy sources — without a solar revolution in the wake of the Industrial Revolution — the Western model of democracy and capitalism is not the perfection of history but its execution.[25]

But those are just a bunch of tree-hugging eco-freaks, you may be thinking. Let's take a look at what some people from the oil industry are saying. Matthew Simmons, a Republican investment banker and member of Vice President Dick Cheney's energy commission, was quoted in June 2004 as saying:

> Oil is far too cheap at the moment [then about $40 a barrel]. The figure I'd use is around $182 a barrel. If we price oil correctly, it could give us time to find bridge fuels, fuels to fill the gap between an oil economy and a renewable economy.[26]

Or perhaps you would prefer the statements of the venerable Deutsche Bank, which wrote in the research report *Energy Prospects After the Petroleum Age* in 2004:

> Venturing to look farther forward on the supply of energy, say to the end of our century, by then the future will already be behind us, at least in terms of petroleum. The end-of-fossil-hydrocarbons scenario is not therefore a doom-and-gloom picture painted by pessimistic end-of-the-world prophets, but a view of scarcity in the coming years and decades that must be taken seriously. Forward-looking politicians, company chiefs and economists should prepare for this in good time, to effect the necessary transitions as smoothly as possible.[27]

With the exception of the small group that happens to be running the United States, everyone agrees that we have to act now. The Fossil Age is a one-time gift that bestowed upon humankind a level of prosperity unimaginable only a few centuries ago. If we now use the remaining resources to set up a large-scale supply of renewable energy, we will be able not only to maintain this level of prosperity but also — and this is unprecedented — to live in harmony with the planet at the same time.

Now let's roll up our sleeves and get down to business.

Chapter 3

Oil

The American way of life is not up for negotiation.
—*President George H.W. Bush at the Rio Earth Summit in 1992*

The End of Oil?

"From now to 2020, there will be no shortage of fossil fuels," the World Energy Council's website read in 2006.[1] I guess that means we will have enough cheap oil for another 14 years. But wait. In September 2002, Robert Priddle, executive director of the International Energy Agency (IEA), stated that the world would have enough fossil resources for the next 30 years.[2] Sounds better! Do I hear 40? Yes, from BP, but unfortunately they say not that we will have enough fossil fuels for the next 40 years but that current reserves will be completely exhausted by then.[3]

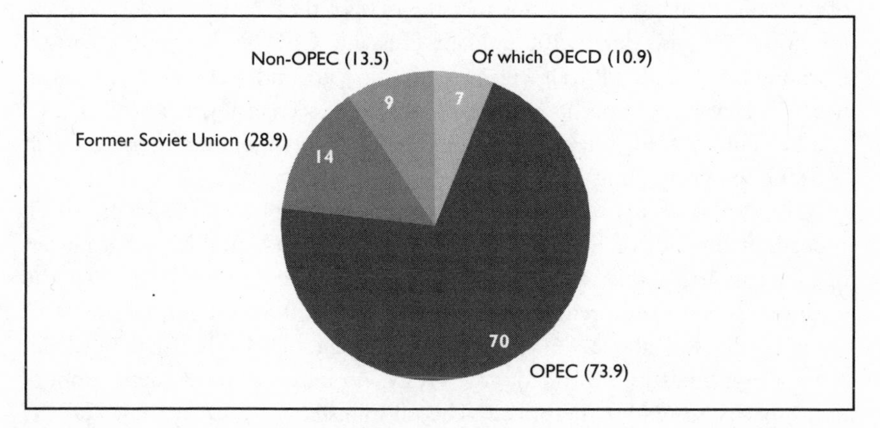

Figure 3.1:
The distribution of global oil reserves. The number of years that current oil reserves will last at current levels of consumption is given in parentheses in the legend. The global total in 2005 was 40.5 years. However, since consumption is increasing, the figures seem to be a complete fabrication. (Based on BP's *Statistical Review of World Energy*, June 2005)

These reserves reflect the amount of oil that can be brought to market using current technology. There is additional oil in the ground but we are going to have to dig deeper to get it, and the fields are getting smaller and smaller. In other words, the oil that is in the ground will be more expensive. And there is no guarantee that we will be able to find more just because prices go up. In 2003, we consumed 28 billion barrels of oil but found only the equivalent of four billion barrels in new fields.[4]

To make things worse, almost all forecasts, such as those of the IEA, expect demand for oil to skyrocket. We should be careful not to interpret charts that show that the demand for oil (unfortunately sometimes labeled "oil consumption") will increase to mean that the supply of oil can meet this demand. The demand for oil may actually increase by 50 percent in the next 25 years but that does not mean production will. As a result, prices — not consumption — will skyrocket.

In 2005, oil prices have remained high and there has been nothing but bad news. Apparently everyone — including Saudi Arabia — is producing full blast. Before his meeting with Saudi Crown Prince Abdullah on April 4, 2005, George W. Bush told the press: "I think they're near capacity, and so we've just got to get a straight answer from the government as to what they think their excess capacity is."[5] White House spokesperson Scott McClellan also did not mince words: "You have to look at the situation in Saudi Arabia as well, and I think most people recognize they are producing at near capacity already."[6]

These are surprisingly clear statements. Are we finally sensitized to the issue of peak oil? Perhaps, for it seems that the days of the $25 barrel may be gone for good. In mid-March 2005, Hani Hussain, CEO at Kuwait Petroleum Corporation, stated: "Prices will never [again] go under the $40 per barrel mark."[7] That same month, Adnan Shihab-Eldin, secretary-general of OPEC, said he could not rule out the possibility that a barrel of oil might cost as much as $80 by the beginning of 2007.[8]

OPEC now seems to have abandoned its target price of $25 per barrel. Indeed, at the end of February 2005 Ali Naimi, Saudi Arabia's oil minister, openly stated that this target had become unrealistic. He predicted that the price of a barrel would remain between $40 and $50 for the rest of the year.[9] (It generally was above $60.) At the beginning of March, Chakib Khelil, Algeria's oil minister, explained why: OPEC no longer had sufficient production capacity to reduce the price.[10] The oil minister of Qatar agreed: "OPEC has done all it can do. This is out of the control of OPEC."[11]

The writing now seems to be on the wall. Even the IEA has stopped beating around the bush. An IEA study with the somewhat disquieting title *Saving Oil in a Hurry*, produced as a basis for discussion at its meeting in May 2005, states unequivocally: "The reality is that oil consumption has caught up with installed crude and refining capacity."[12]

In other words, not only is oil being pumped out of the ground at full capacity but also refineries are running full blast. In addition, though the IEA does not mention it, even if we could get a substantial amount of additional oil out of the ground there is basically no tanker capacity left. Interestingly, it seems that the entire production chain in the oil industry is ready for oil to peak right at this level. As British energy expert Jeremy Leggett put it in 2005:

> Goldman Sachs has calculated that the oil industry needs to invest $2.4 trillion over the next 10 years to do the exploration and put the infrastructure in place to meet projected global demand. The oil industry, vastly rich as it is, has never come close to such sums in the past. This is nearly three times the level of 1990s spending. Do the oil companies really believe there is enough oil out there to justify such investment? We will see, for sure, in the next few years.[13]

On April 7, 2005, the International Monetary Fund spoke of an impending "permanent oil shock" and added that the world would have to get used to higher oil prices.[14] And Chris Skrebowski, editor of the British *Petroleum Review*, stated that "new production would almost certainly not be sufficient to offset diminishing supplies from existing sources and still meet growing global demand." He expects production to increase from 82.5 million barrels a day in 2004 to 85 million by 2007-2008. After that, he believes production will drop once and for all.[15]

A number of investment banks also expect the price of oil to increase. For instance, CIBC World Markets predicted on April 13, 2005 that a barrel of oil could cost as much as $100 by 2010. Unlike Skrebowski, CIBC believes that as much as 86.8 million barrels can be produced per day but also predicts that this increase will fall far short of the demand for 95.7 million barrels by that time.[16] And at the end of March 2005, Goldman Sachs, one of the largest investment banks in the energy sector, forecast that a barrel could cost as much as $105 in the short term.[17] What would happen after that? In April 2005, French investment bank IXIS-CIB warned that a barrel of oil could cost $380 by 2015 if nothing changes.[18]

What could change? As we saw in the previous chapter, neoliberal economics holds that higher prices will produce greater supply — or, as George W. Bush put it in 2005, "I will tell you with $55 oil, we don't need incentives to oil and gas companies to explore."[19] What Bush does not tell us is that, as *The New York Times* reported on March 25, 2005, twice as much was invested in oil exploration in 2003 as was found.[20]

What about technological advances making some of the oil we have already found profitable as market prices increase? Hopefully, there is much potential here but the risks are even now making themselves felt. For instance, for many years a variety of substances ranging from water to air, nitrogen, and more recently carbon dioxide have been injected into oil wells to maintain production levels by increasing the pressure below ground. We now know that this method can greatly foreshorten the productive life of a field without notice.

Another method, called horizontal drilling, can have the same detrimental effect. In this approach, one vertical hole is drilled and from a certain point drilling then continues horizontally underground to reach other parts of the oilfield. This method offers two benefits. From an environmental perspective, it is beneficial to have fewer boreholes at the surface, especially in such areas as the Arctic National Wildlife Refuge (ANWR). In addition, if the oilfield is very deep it often is more economical to drill one hole several kilometers deep and continue drilling horizontally rather then have multiple vertical boreholes several kilometers deep.

But as Shell discovered in its largest field (Yibal in Oman), horizontal drilling also can damage a field. At the beginning of 2004, Shell reduced its reported reserves by 20 percent, mostly because of damage at Yibal caused by horizontal drilling. At the beginning of 2005, Mexico announced that its largest oil field (Cantarell) — the second largest in the world in terms of production volume — was suddenly not producing as much. At this field, nitrogen is injected to keep up production levels. Normally the field would produce as much as Kuwait. Now levels seem to be dropping fast.

Currently, everyone is looking toward Saudi Arabia. Can it increase its production? In his 2005 book *Twilight in the Desert: The Coming Saudi Oil Shock and the World Economy,*[21] Matthew Simmons claims that the Saudis have damaged the world's largest oilfield (Ghawar) by desperately trying to keep production levels up. Canadian investment bank Harris Investment Management, Inc. expressed the same concern in its *Basic Points* report of March 2005, "Big Footprints on the Sands of Time and Little Footprints of Fear."[22]

The experts do not agree on what the consequences of peak oil will be. Optimists hold that the end of cheap oil will be a good thing for the climate and the environment and that we can easily make the transition to alternative energy. But what exactly are we going to replace oil with? Electricity from windmills and solar panels? Most experts are skeptical. As the February 2005 report from the US Department of Energy *Peaking of World Oil Production: Impacts, Mitigation, & Risk Management* puts it:

> The world has never faced a problem like this. Without massive mitigation more than a decade before the fact, the problem will be pervasive and will not be temporary. Previous energy transitions (wood to coal and coal to oil) were gradual and evolutionary; oil peaking will be abrupt and revolutionary.[23]

We therefore must hope that OPEC will at least be able to keep the price of a barrel relatively stable for a number of years even if it is not able to lower the price. Otherwise we might have to implement the proposals made by the IEA in *Saving Oil in a Hurry:* speed limits of 55 miles per hour will have to be imposed on private cars if we are going to keep them at all, and telecommuting will be favored over a shortened work week.[24] Perhaps the motto for the 21st century should be "Data highways for pedestrians."

These predictions are not unrealistic warnings from radical environmental activists but the informed opinions of oil experts. I was most surprised when Texas oil tycoon T. Boone Pickens said in 2004: "Never again will we pump more than 82 million barrels."[25] Whether he is right or not, his statement is especially important because so many scenarios assume that production can be increased to 100 million barrels a day.

According to the BP *Statistical Review of World Energy* of 2005, world oil reserves will last another 40.5 years, natural gas 66.7 years, and coal 164 years.[26] All these figures assume that consumption will not increase. But even if these figures are off by a couple of decades one way or the other, the Fossil Age will appear as a fleeting moment in the history of civilization: 300 years from around 1850 to 2150. Ever since oil was discovered in Pennsylvania in 1859, we have been running out of it, burning it far faster than it can regenerate.

In the past few years, opponents of the preemptive strike against Iraq have argued that this was a war for oil. Keep in mind that while Iraq has almost 11 percent of current reserves at 112 billion barrels — more than any other country except Saudi Arabia — the world currently consumes some 26 billion barrels every year. The oil in Iraq alone would not cover the world's thirst for oil for even five years. The United States doesn't have 30 billion barrels left in its own territory. For some 20 years, oil production has exceeded discoveries in the ground worldwide.[27]

We begin to see why the Bush administration is so intent on tapping the estimated 16 billion barrels in the Arctic National Wildlife Refuge in Alaska. But even if we can get all 16 billion barrels out of Alaska, which most experts doubt, that amount would be a drop in the bucket. Unfortunately, the press is not doing a good job of explaining this. Reuters stated: "Americans currently consume about 20 million barrels of oil each day and the Arctic refuge could contain as much as 16 billion barrels of oil."[28] Too bad Reuters does not go on to explain what that equation means: the Arctic reserves would cover US consumption for only two years. And if we take the lower estimate of the Natural Resources Defense Council, we do not have 16 billion barrels of oil in ANWR but only 3.2 billion.[29]

But haven't there always been people claiming that we were going to run out of oil soon? Yes, as J. Howard Pew, then president of Sun Oil, explained way back in 1925: "Periodically ever since I was a small boy, there has been an agitation predicting oil shortage, and always in the succeeding years the production has been greater than ever before." [30] Keep in mind, though, that in 1925 no one had even thought of looking for oil in the Persian Gulf. In other words, most of the planet's oil had yet to be looked for, much less found.

Today, on the other hand, the whole planet has been explored and no researchers think there are any giant oil fields waiting to be discovered. In 2000, the US Energy Information Administration warned people to lower their expectations: "The number of frontier areas is diminishing." [31] Geologists speak of an "oil window" within which oil can be found. We will not be able to simply drill deeper and deeper because at a certain depth (around three kilometers) we start getting less and less oil and more gas until we eventually run out of both.

An End to "Easy Oil"

Authors often do not choose the titles for their own books. Instead, the publisher gives the book a title that it hopes will sell. Sometimes the title does not match the book's message very well. Quite possibly this is what happened with Paul Roberts's recent book *The End of Oil*.[32] True, Roberts is concerned about the impending end of the hydrocarbon industry, but he also takes great pains to point out that he is talking about the end of "easy oil," not all oil. What difference does that make?

Basically, the oil left in the ground will not be cheap, nor will we be able to bring it to market in amounts large enough to keep up with demand. While Canada may have oil reserves almost as great as Saudi Arabia's, almost all of it is locked up in oil sands. It may be possible for Canada to bring a small amount of this to market for many centuries to come, but it will never be able to match the current production levels of Saudi Arabia.

Roberts does not believe that the market will take care of this problem. He points out that when oil becomes more expensive everything else will too: not just transportation but also the products we transport — everything from strawberries to PCs — and, of course, all products made of oil (such as plastics and fertilizers). He believes that if we do not prepare for peak oil the crisis that will ensue will make the Great Depression look like a "dress rehearsal." Roberts correctly points out that we will not know when peak oil has been reached until after the fact. But he argues that the conflict in Iraq suggests that the battle for the oil that remains already has begun.

At present, every country — with the possible exception of Saudi Arabia — is producing oil at maximum capacity. No one else can increase production and the Saudis cannot increase it enough to keep up with demand. In other words, the Saudis can no longer control the price of a barrel of oil. In his book, Roberts asks a Saudi official whether there are any disagreements between Saudi Arabia and Russia, which has been increasing its production recently to levels above what was agreed with Saudi Arabia. The official answers, in the tone typical of Saudi oil men in the past few decades, that the Saudis could flood the market with oil at any time and wipe out Russian investments if they wanted to. Just one year after Roberts's book was published, the Saudis no longer seemed able to flood the market.

The Market is Not Free

Brent oil, which is oil from the North Sea, is some of the most expensive oil currently brought to market. It is so expensive that it would have been completely uncompetitive before prices skyrocketed in the two artificial bottlenecks of the 1970s. Since it costs at least $10 to pump a barrel of oil out of the North Sea, if the price of a barrel on the world market drops below $15 the whole North Sea enterprise threatens to become unprofitable. In contrast, the oil from Saudi Arabia is some of the cheapest in the world, costing only around $2 a barrel to pump. The only country in the world that can compete with those prices on a large scale is Iraq.

Until recently, successive US administrations and Saudi Arabia have been trying to keep the price of a barrel of oil at around $25. At this level, investments in the rather expensive offshore projects — such as in the North Sea, off the coast of Africa, and in the Gulf of Mexico — remain competitive. If prices rise too high, the oil industry fears that energy efficiency and alternative sources of energy will be developed to the detriment of oil. As a result, the Oil Age might end before we run out of oil completely. As none other than Saudi oil minister Zaki Yamani famously put it in 1974: "The Stone Age didn't end for lack of stone, and the Oil Age will end long before the world runs out of oil."

What all this means is that the price of oil has been kept artificially high so that a certain amount of expensive oil could compete with cheaper oil from the Persian Gulf region. In other words, for decades we have been producing more oil than normal market prices would have allowed. And in the past ten years we have been consuming roughly two and a half times as much oil as companies have discovered through exploration.

We now begin to see why Roberts speaks of the end of "easy oil." To begin with, Saudi Arabia and Iraq, for instance, probably can continue to produce oil in large quantities for many decades to come, whereas other OPEC countries will be running short soon, probably by 2020. Ironically, it seems that we are running out of the expensive oil before we run out of the cheap oil.

But is Yamani correct? Can we really do without oil? According to many estimates, the United States certainly could. It would be easy enough to cut consumption in half if Americans would simply start driving more efficient cars such as many of the models sold in Europe that are not even available on the US market. Late in 2004, Amory Lovins and his team at the Rocky Mountain Institute published *Winning the Oil Endgame,* which set forth a plan to cut oil consumption by 50 percent over the next 20 years. According to their estimates, there would even be a net benefit to the economy, with the costs avoided exceeding the investments at the outset by a wide margin.[33]

When will the United States go down this road? For decades, Amory and Hunter Lovins have been pointing out all the ways we can save money, decrease our dependency on energy imports, and help the environment. In January 1995, they wrote in *The Atlantic:* "We Americans recently put our sons and daughters in 0.56 mpg tanks and 17-feet-per-gallon aircraft carriers because we hadn't put them in 32 mpg cars."[34]

Some ten years later, Roberts described his fellow Americans as babies who cry when you take away their pacifier. We truly are addicted to oil and completely in denial. We're not even aware of how much energy we consume. Most of us probably have a rough idea how much we pay for electricity a month but not how much using a particular appliance costs us. (Do you know how much electricity you consume when you toast your bread in the morning?) As Roberts says, we are energy-illiterate.

The first step therefore must be to raise awareness of how we waste energy. Roberts's book is a good step in this direction, as are the Lovinses' works. But these authors believe that the hydrogen economy is somehow going to save us, though they do not know where we are going to get the hydrogen (see Chapter 11). To my chagrin, Roberts does not believe that renewables can be the solution:

> On average, analysts say, wind and solar renewables can provide a maximum of 20 percent of a region's power. Past that point, either the intermittency factor causes too many power disruptions, or the cost of maintaining so much backup base load becomes too high.[35]

Although Roberts repeatedly mentions my current hometown of Freiburg in southeastern Germany, he does not seem to be aware that wind power alone already covers 20 percent or more of the power supply in Denmark and parts of northern Germany. Furthermore, a recent study by the German Energy Agency (dena) found that the share of renewables in the grid could be greatly increased.[36] After all, when solar power is added on a grand scale we will have more power at times of peak consumption. Other sources of renewable energy such as biomass and geothermal plants can be switched on as needed just as gas turbines are today. Some analysts therefore believe that renewables will be able to provide 100 percent of our electricity supply.

Roberts thus sounds a bit like US Vice President Dick Cheney when he says that while we need technologies that generate energy without emissions these technologies either are not ready for the market or have not been invented yet. My purpose is to prove the exact opposite and to show that Germany is leading the way in the transition to renewables. For while I disagree with Roberts on this point, we agree on one thing. As he puts it: "Ultimately, the question facing us isn't whether our energy systems will change — indeed, the process is already well under way — but whether we can live with the outcome."[37]

Esso Declares 2003 an "Oil Dorado"

While European oil firms such as Shell and BP are adding renewables to their portfolios, American oil firms such as ExxonMobil are sticking to petroleum. No wonder that Esso, the European subsidiary of ExxonMobil, would have us believe in its "Oil Dorado" (*Öldorado 2003*) report that we suddenly had a lot more oil in 2003 than in 2002. The trick is that Canada turned much of its "resources" into "reserves," the latter being resources that can be brought to market at competitive prices. According to the Esso report, in 2003 Canada had the second-largest oil reserves in the world behind Saudi Arabia.[38]

However, Canada's oil sands cannot be compared to the petroleum from the Persian Gulf. While it only costs a few dollars a barrel to produce oil in Saudi Arabia and Iraq, for instance, producing a barrel of oil from oil sands can easily cost 15 to 20 times as much. In 2002, it cost $40 to produce a barrel of oil from oil sands in Canada, though this figure is dropping and may have reached $30 in some projects.

Does this drop not indicate that oil sands are now profitable, thus justifying their inclusion as oil reserves? In light of oil prices around $60 a barrel, it would indeed seem that the days of profitable oil from oil sands has finally come. But not all experts are convinced. In an interview for this book, Jörg Schindler, chairman of the Global Challenges Network, described the reporting of Canadian oil sands as oil reserves as a "cheap trick." Schindler pointed out that oil has been produced from these sands in Canada since the 1960s, but until recently this process was considered too expensive to be competitive. Only 10 to 20 percent of the oil sands in Canada consist of oil, the rest being mostly sand and a bit of water. Not only does the sand need to be washed out to extract the oil but also the oil thus extracted needs to be cooked in an energy-intensive process to produce the quality of oil required by the market.[39]

Schindler and his colleague Werner Zittel, coauthors with Colin Campbell and Frauke Liesenborghs of the German book *Ölwechsel!* on future world oil production,[40] point out that a quarter of freshwater consumption in Alberta, Canada, already is devoted to oil production from these sands. In other words, it will not be possible to increase production very much, at least not without disastrous environmental consequences.

So while Esso's 2003 annual report may have increased Canada's reported oil reserves from 654 million tons in 2001 to 24,235 million tons in 2002, Canada will not be able to play a major role alongside Saudi Arabia's 35,409 million tons or Iraq's 15,095 million tons because Canada will never be able to produce 10 million barrels a day as Saudi Arabia already has done.

Other problems with Esso's 2003 report are indicative of the basic problem with all reported oil figures. The increase in global oil reserves is clearly the result of the reassessment of Canada's oil sands. Otherwise, reserves remained basically constant worldwide even though a lot of oil was consumed. Theoretically, a stable volume of reserves is possible if, by chance, the amount of new discoveries and resources reassessed as reserves is equal to the amount consumed. The problem is that oil-producing countries and oil companies apparently never have produced reliable figures.

Concerns about manipulated figures for reported oil reserves are twofold. On the one hand, people are worried that reported oil reserves may be exaggerated, in which case less oil would be in the ground than we currently believe. Here the events of the 1980s often are cited: OPEC countries suddenly increased their reported reserves drastically because the amount a country could produce within OPEC agreements was based on the amount of reserves that country had. If you wanted to produce and sell more, you had to have more in the ground.

On the other hand, perhaps the figures are not that exaggerated, for there is a temptation to do the exact opposite as well since oil companies pay lower taxes when they report lower reserves. In addition, if an oil company underestimates its reserves, it can gradually add on a small amount to keep its reserves basically stable and argue convincingly that better technology is allowing it to produce more at market prices. In this case, there is more oil in the ground than reported.

In either case, there seems to be reason to believe that the figures are being manipulated. The critics of current reporting practices in the oil industry include such insiders as Colin Campbell and Matt Simmons, an adviser to

George W. Bush. Simmons, who predicts in *Twilight in the Desert* that Saudi Arabia has been overreporting its production potential, also points out that 68 percent of the world's oil-producing countries have not changed their figures for reserves since 2002, with 39 percent remaining unchanged since 1998 and 13 percent since 1993.[41] Campbell, who used to work in exploration for oil companies, told the British daily *The Guardian* in April 2005 that, when working for the industry, "I do not think that I ever told the truth about the size of a prospect."[42]

The figures reported by oil firms have not, in other words, been independently verified. Indeed, BP does not itself research the figures it publishes in its *Statistical Review of World Energy* but simply reprints the numbers published by the renowned *The Oil & Gas Journal.* If you think the statistics published by the main oil industry journal must be reliable, think again: the editors of that journal do not corroborate the unchanged figures that companies report year after year.

Former oil exploration experts such as Simmons are now working for investment banks, which do not like risks. Perhaps this helps explain the increase in criticism of current oil reporting standards. As Simmons put it in February 2005, "If we fail to reform [the reporting standards for oil statistics], we deserve any unfortunate surprises that befall us."[43] Indeed, unlike financial reporting, for which there are international standards, the standards for the reporting of oil reserves vary not only from country to country but also from company to company. In other words, no one really knows what anyone else is talking about.

Recently, however, to reduce investment risk the Securities Exchange Commission (SEC) formulated its own standards for reports of oil reserves.[44] The SEC's standards produced much lower figures than the internal calculations of oil firms. For example, the world's largest oil company, ExxonMobil, claimed that it had produced new reserves in 2004 equal to 112 percent of the oil it sold that year; at BP, the internal figure was 110 percent. Not only did the basis for BP's calculations differ from ExxonMobil's but also, when the SEC's standards were applied, BP was able to cover only 89 percent of its production in 2004, with ExxonMobil faring even worse at 83 percent. And if you think those figures sound bad you probably don't even want to hear that Shell has been able to replace only between 44 and 57 percent of its production over the past five years.

Simmons also reminds us that "proven reserves" are not by any means proven and should actually be called "estimated reserves." In other words, the figures reported for oil reserves not only have not been verified by any independent agency but also are inherently unstable. The US Department of Energy's *Peaking of World Oil Production* described this instability:

> Reserves estimation is a bit like a blindfolded person trying to judge what the whole elephant looks like from touching it in just a few places. It is not like counting cars in a parking lot, where all the cars are in full view.[45]

Even given the difficulty of assessing the size of an underground oil field, one can only marvel at the nonchalance with which oil companies produce obviously inaccurate figures. Take, for example, Esso's "Oil Dorado" report, which states that global oil production dropped from 3,585.6 million tons in 2001 to 3,552.4 million tons in 2002. Esso explains the 33.2-million-ton drop as follows: "The main reasons are the weak economy in some parts of the world and OPEC's constant production discipline."[46] What does a weak economy have to do with the amount of oil produced? One would expect a weak economy to reduce consumption but not necessarily production, especially when Esso reports that production *capacity* actually increased in the same period. But consumption actually rose by approximately half a percent. And we don't have to look far to discover what is meant by OPEC's "discipline." As *The Washington Post* reported in the summer of 2002:

> Saudi Arabia has quietly taken steps to protect or even increase its already dominant influence in the world oil market in the face of growing uncertainty about the effect of a war in Iraq on global energy supplies and prices, according to U.S. and Middle Eastern officials and experts. Saudi Arabia has reclaimed its position as the No. 1 foreign supplier of crude oil to the United States in recent months and offered to further increase sales in December, the Energy Department reported. To keep competitors from taking away customers, the Saudis have boosted production by an estimated 1 million barrels a day above the quota set by the Organization of the Petroleum Exporting Countries, according to a New York industry analyst.[47]

And what are we to make of Esso's statement that oil production exceeded consumption by some 9.4 million tons? We would expect oil in stock to have increased, but Esso writes: "Global oil consumption increased by around 16 million tons to 3,543 million tons. In combination with the drop in production, worldwide stocks of oil were noticeably reduced." What happened to the 9.4 million tons of oil that were produced but not consumed? Esso does not tell us. Maybe what Esso means is that the expansion of oil reserves was reduced. But in light of all of these inconsistencies, one can only be skeptical about all Esso's figures.

Interestingly, the Esso study not only contradicts itself but also is not supported by BP's *Statistical Review of World Energy* of 2003, published seven days before Esso's study. Did the Canadians wait for BP to publish its study so that Esso could have the exclusive story? Whatever the case, in 2003 BP reported that Canada had around 900 million tons of oil — not almost 25,000 million. However, BP's 2003 report is interesting also because in many cases the figures for oil reserves did not change. One can imagine that Iraq, whose oil production had been sanctioned, still had 112.5 billion barrels but it is truly amazing that no other OPEC country indicated it had lower reserves in 2002 than in the previous year. In fact, the figures changed in only two countries: Ecuador (an increase from 2.1 to 4.6 billion barrels, more than twice as much) and Venezuela (a minor increase from 77.7 to 77.8 billion barrels). The following OPEC countries did not change their figures at all from 2001 to 2002: Iran, Saudi Arabia, Kuwait, Qatar, United Arab Emirates, and Indonesia. In addition, no African country — including Nigeria, Libya, Gabon, Angola, and Algeria, to name just a few of the largest oil producers in Africa — changed its reserves at all. Apparently, all these countries send the same letter every year to *The Oil & Gas Journal,* with the only difference being the date.

BP's figures seem to suggest that the oil peak is moving further and further away. In 2003, BP's *Review* estimated that proven reserves would last another 40.6 years at the current rate of consumption. The figure the previous year had been 40.3 years. By 2004, the figure had risen to 41.0 years. In 2005, the figure remained above the level of 2002 at 40.5. What is the layman to make of all this?

BP's 2004 report is also interesting because the figures for Canada had been changed retroactively. In the 2003 *Review,* Canada had 900 million tons (at the end of 2002), but in the 2004 *Review* it suddenly had 2,300 million tons. No problem, you may be thinking: BP finally closed ranks with Esso and changed Canada's reserves accordingly. You might be right, but this doesn't explain why the 2004 *Review* claims that Canada's reserves were 4 percent lower in 2003 than in 2002. In other words, BP's figures for Canada's oil reserves for the years up to and including 2002 do not correspond to those reported by Esso or even to those from BP's 2003 *Review.* Unfortunately, the 2004 *Review* does not provide any explanation of these changes.

One thing seems clear: cheap oil is increasingly being replaced by expensive alternatives that are gradually becoming marketable because oil prices are rising. So even if we believe the figures that BP and Esso publish, we may have lots of oil left, but it ain't gonna be cheap. Or as Shell Nederland president Rein Willems put it in an interview with the Dutch journal *De Ingenieur* in 2005,[48] Shell expects oil prices to increase because production costs are rising (i.e., cheap oil is running out), though he was quick to add that Shell believes "there is enough oil to last a hundred years." Shell, he explained, wants to invest in renewables because of climate change, not peak oil. Feel better now?

Chapter 4

Biomass

Bioenergies are the all-rounders among the renewables, since only biomass is equally suitable for the generation of electricity and heat as well as the production of fuels .... Should the oil price rise to USD 100/barrel and beyond, biofuels would become competitive, even at the current technological stage.

—Josef Auer of Deutsche Bank Research, 2005

A Source of Energy or Food?

Biomass is the classic source of renewable energy. From the Stone Age until around 1800, when the Fossil Age began with coal, wood was by far the number one source of energy. Even today, biomass makes up the largest share of the total energy supply of the US. We are rightly impressed to hear that Denmark already gets around a fifth of its electricity from wind, but according to the IEA[1] biomass provides a full 14 percent of the global supply of *energy* (not just electricity). In the EU, biomass makes up two-thirds of total primary energy from renewable sources. In addition to conventional biomass, which entails problematic environmental and health impacts, "new biomass" is now often used in technologies such as pellet-fired stoves (especially in Austria) and fuel cells or as an alternative fuel (mainly ethanol in the US and rapeseed oil in the EU) to gasoline and diesel.

What is the potential of biomass? Do we have enough arable land to grow energy plants in addition to food? And if we take all the fields currently not being used for agriculture to grow some "super plant," will we not have monoculture as far as the eye can see?

The IEA states that biomass currently makes up 80 percent of renewables worldwide. It tends to be used more in poor countries. In the developing world it provides around 25 percent of the total energy supply, with outliers up to 90 percent, such as Uganda, where only 3 percent of the original natural forest is left standing.[2] Such conventional biomass not only threatens to be unsustainable when certain plants are overharvested, causing potentially irreversible environmental damage, but also can be hazardous to human health when people breathe in the smoke from open fires.

"New biomass" replaces fossil energy carriers in modern technologies and thus is at least potentially more sustainable, more environmentally friendly, and less hazardous to human health than conventional biomass. In the end, however, the shift from conventional biomass to new biomass probably will not increase the share of biomass in global energy consumption but only keep it stable. According to a scenario of the IEA,[3] the share of biomass could even drop from 14.2 percent in 2000 to 11 percent by 2020 if biomass in combination with new technologies replaces conventional biomass.

Nonetheless, biomass will play an important role in a future mix of renewables because, unlike wind power and solar power, biomass can be turned on and off to cover fluctuations in demand. The only other source of renewable energy that can do the same is geothermal power, but unlike biomass geothermal cannot be used as a fuel in mobile applications. In other words, this new biomass has a unique role to play in a renewable energy supply: it will be available all year long, day and night, at the touch of a button and it can be used for heating, to generate electricity, and for motive power.

The Potential of "New Biomass"

Again and again we read that we are running out of oil and should therefore step up installation of wind turbines and solar panels. We need to remember, however, that wind turbines and solar panels make electricity, and electricity cannot replace oil. Biomass can. It can even be used to manufacture products such as plastics. Furthermore, some biomass is basically waste products, from manure to vegetable oil that has already been used in deep fryers. Indeed, cars that smell like french fries have become somewhat of a cult item among environmental activists in the US.

Before petroleum was used on a grand scale, vegetable oils were used for lubrication and fuel. Remember that the first diesel motor ran on peanut oil. Today, the use of rapeseed (canola) oil is being stepped up in Germany: since the beginning of 2004, all diesel sold in Germany has consisted of 5 percent rapeseed oil. Of course, since combustion engines have been tailored to fossil fuels for more than 100 years, switching to 100 percent biofuels would mean that motors might not run smoothly, but blends generally are not problematic. In fact, in Colombia and China, gasoline contains 10 percent ethanol; in Brazil, up to 26 percent.

According to EU directive 2003/30/EEC,[4] by the end of 2010 biofuels will have to make up 5.75 percent of the EU's fuel consumption. Industrialized countries lag behind developing nations and newly industrialized countries when it comes to biomass consumption but the biofuel markets in Europe are booming, with German production of biodiesel increasing in 2004 by approximately 25 percent over the level of 2003 to around 1.2 million tons.

The World Energy Council even calls biomass "potentially the world's largest and most sustainable energy source."[5] But the debate about the pros and cons of biomass really gets hot only when we start talking about large-scale production. How much land do we need to grow rapeseed if we are going to replace diesel completely with rapeseed oil? In Germany, where some 40 percent of private cars are diesels, it has been estimated that two-thirds of the country would have to be turned into fields of rapeseed.

Calculations for other countries have come to similar findings. For instance, British environmental journalist George Monbiot of *The Guardian* has estimated that all of England would have to be turned into a rapeseed field if the United Kingdom (Great Britain and Northern Ireland) had to make do with rapeseed oil for all of its fuel needs.[6] British researcher Jeremy Woods of Imperial College London found that, to cover a mere 5 percent of the energy needed for transport, between 12 percent (sugar beets) and 45 percent (wheat straw) of arable land would have to be devoted to energy plants.[7] It would seem that at current rates of consumption the UK cannot possibly grow its own transport energy.

Aside from the fact that it would not be possible to cover England with fields of rapeseed — people need not only space to live but also space to grow food — such a monoculture would be a disaster for the environment. While biomass is CO_2-neutral (the same amount of CO_2 is emitted as the plant absorbed) and is thus more environmentally friendly than petroleum, large-scale plantations of one type of plant would drain the soil of particular nutrients and devastate the diversity of animals and insects.

How much energy can we get from biomass sustainably? Estimates for global potential vary greatly from less than 10 percent to more than 300 percent of our current energy consumption. A Dutch research team from the University of Utrecht reviewed 17 international studies on the global potential of biomass. Some of the studies postulated that biomass will be able to cover up to 50 percent of global energy consumption by between 2050 and 2100.[8]

How much of the current 11.8 million hectares of agricultural land can be used to grow energy plants instead of food? In a 2004 report for the German Ministry of the Environment,[9] Germany's Institute of Applied Ecology found that better harvests could make some 4.4 million hectares of land currently used in agriculture available by 2030 for energy plants thanks to improved technology and Germany's shrinking population. By then, biomass would be covering some 25 percent of the country's total energy consumption. What is especially interesting in this scenario is that organic farming would increase to

20 percent and monoculture would be prevented. In other words, biomass would remain environmentally friendly and sustainable.

On the other hand, another study by the German Ministry of the Environment[10] found the short-term potential much more modest. By 2010, only 0.2 million additional hectares would become available for the sustainable cultivation of energy plants, partly because Germany's aging population is not expected to start shrinking until around 2020.

Since population is not shrinking everywhere, what can we say globally about the potential of biomass? Some prominent proponents of renewables believe that biomass will be able to provide a much larger share of our energy sustainably. For instance, Hermann Scheer has written: "Some 2 million km of forests can be sustainably managed to provide enough biofuel to cover the world's current level of consumption."[11]

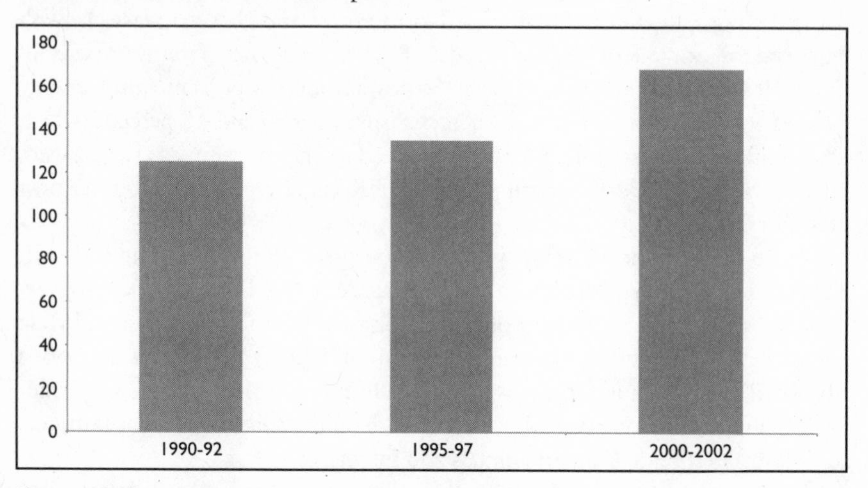

Figure 4.1:
The 2001 net energy balance of corn ethanol. Technology not only is providing greater agricultural yields but also is improving the conversion of biomass into final energy. Between 1990 and 2002, the energy balance of ethanol made from corn increased by around 33 percent in the US. (Source: Hosein Shapouri, USDA, PowerPoint presentation available on the web)

How much space is that? According to the United Nations Environmental Program (UNEP), the world currently has approximately 45 million square kilometers of forests.[12] The German Association of Energy Consumers explains that:

> the amount of energy that regenerates in the planet's forests each year is 25 times greater than global oil production at approximately 3.5 billion tons annually. Five percent of the world's forests would be enough to cover global oil consumption. To do so, we would

> not need to plant any energy crops or recultivate any semi-arid regions, which together make up an additional 49 million square kilometers.[13]

Can we get a third of our energy supply from gasified wood? Perhaps, but there is reason to be skeptical about whether such a large-scale operation would be truly sustainable. After all, countries like Haiti are practically denuded and unsustainable forestry already is a major problem in countries from Brazil to Indonesia. When the Philippines was flooded at the end of 2004, the world saw what can happen when heavy precipitation hits denuded forest hills. In addition, as *Le Monde diplomatique* reported in 2004,[14] forestry already is as corrupt an industry as the oil sector, and disputes about these resources have led to violent conflicts in Cambodia, Sierra Leone, Ivory Coast, Congo, Burma, and Liberia.

The United States (200 million hectares, roughly a fourth of its total area) and Canada (400 million hectares, some 40 percent of the country) have massive forests at their disposal but the EU is not as fortunate and will eventually have to rely on imports. The question is whether other countries will sell their biomass on the markets and at what price. Western Europe may be able to become less dependent on oil imports only by switching to imports of biomass. Densely populated industrialized countries like Germany have very high energy consumption per hectare. While per capita energy consumption in the United States and Canada is about twice as great as in the EU, consumption per hectare is much less because North America is so sparsely populated. Europe has the lowest woody biomass density in the world at 59 tons per hectare, far below the world average of 139 tons per hectare.[15] According to the Food and Agriculture Organization of the United Nations, North and Central America have 104 tons of woody biomass per hectare.[16]

We also should keep in mind that the energy in the plants we feed to farm animals is not completely lost. One cow produces from 100 to 200 liters of methane each day. California, home to two million cows, is already taking steps to recapture this energy. Currently, there is a lot of investment in the gasification of biomass, and biowaste is already being used in fuel cells. In Germany, biogas is being tested as a fuel in PEM fuel cells (see Chapter 11).

Generally, biomass has been prepared for use as an energy carrier by means of pyrolysis: the biomass is heated in an oxygen-free environment to produce an energy carrier with a greater calorific value. (Charcoal is a traditional form of biomass that has undergone pyrolysis.) But new methods of gasification are being developed. For instance, in pyrolysis the coke that is created as a byproduct can be gasified with the addition of air, but if water vapor is added hydrogen is created instead of CO_2. The German company H2Herten is doing just that in what it calls its Blue Tower.[17]

This Blue Tower is interesting because the process can be used on any type of biomass. If biomass is to be used sustainably, we will have to go beyond looking for a "super plant" and also begin using all of the plant, not just the oil-rich seeds. The energy balance of a particular energy carrier indicates how much energy the carrier provides minus the energy required to produce it. One can easily be tempted to try to find out which plant has the greatest "energy return on investment" (EROI). But, as we will see, this figure is not always easy to determine and it also makes sense to take advantage of the energy in some waste products with rather poor energy balances when such products are available anyway.

THE ENERGY BALANCE OF BIOMASS

Before we discuss which plants have the best energy balance, let us first take a critical look at the concept. Biomass is often criticized because its energy balance is so poor. This is true in comparison to other renewables.

Energy source	Energy balance
Wind	10-20
PV	> 5
Ethanol from wood	> 2.5
Ethanol from corn	> 1.6

Figure 4.2:
Energy balance of some renewable energy sources. Over its service life, a wind turbine generates 10 to 20 times the amount of energy required for its manufacture, operation, and dismantling. Photovoltaics easily produces five times the amount of energy it requires. But it's a close call with ethanol; some even claim that in some cases more energy is required to make ethanol from corn than is provided in the end. (Source: Rounded figures based on various studies)

In addition, it is often claimed that within a few months or less a coal-fired plant or a nuclear power plant produces as much energy as was invested in the plant, while a wind turbine takes almost a year to do so and photovoltaics takes several years. This is true only if we forget that coal plants and nuclear plants devour tremendous quantities of non-renewable resources. These finite resources are never subtracted from the equation.

To see what a difference this makes, let's start with a typical coal plant with an efficiency of around 35 percent. Roughly one-third of the energy in the coal is converted into electricity, with almost two-thirds exiting the plant's cooling towers as waste heat. In other words, you always put three times as much energy into a coal plant as you get out. Once we take account of this fact, the energy balance of such a plant is -3.

In addition, the energy balance of oil has been worsening drastically in recent years. According to a report in *Nature* in November 2003 titled "Hydrocarbons and the evolution of human culture,"[18] we currently get 17 times more energy from the oil we produce than we invest in its production. But 100 years ago, the energy return on investment was around 100 for oil. The remaining oil fields simply do not produce large amounts of oil easily, and Canada's oil sands have a miserable energy balance of −3 or possibly lower.

If we apply the same logic to photovoltaics, which generally has efficiencies of 10 to 15 percent, then we also have to rate its energy balance at −6 to −10. But we do not need to do so because no energy is ever wasted with photovoltaics. If we double the number of coal plants, we will run out of coal much faster. But if we put solar panels on every house, the sun is not going to stop shining any sooner. In other words, what we should be measuring in addition to the efficiency with which a solar cell converts sunlight into electricity is the amount of solar radiation incident on the Earth every day that we turn into electricity.

One could argue that simply allowing sunlight to shine on a black-shingled roof is actually wasting the sunlight, which could have been converted into electricity. We therefore should be measuring what I would call our energy "macrobalance." We have a certain amount of sunlight, wind, tides, geothermal energy, and such at our disposal every year whether we use it or not. We should measure how much of this we use as final energy and strive to increase this figure quickly and consistently.

Our macrobalance would also be improved if we increased the use of biomass for non-energy purposes. For instance, when power lines are run across wooden poles instead of concrete or steel pylons, the CO_2 emissions from the manufacture of the pylons is reduced considerably.

Pylons made of	Tons of CO2 per kilometer of power line
Wood	4
Concrete	17
Steel	38

Figure 4.3:
CO_2 emissions from pylon manufacture (Source: IEA Bioenergy)

Biomass binds CO_2 and other elements during plant growth. These elements are emitted when the plants rot or are burned, but the amount is always equal to the amount the plant absorbed during its growth. The same is true of fossil fuels, with the difference that we burn not roughly the same amount of fossil energy as regenerates every year but rather around 15,000 times as much. With *sustainable* biomass, we consume about the amount that grows back each year. A comparison of the energy balance of renewables with the energy balance of finite energy sources thus is completely misleading. Once these finite resources have become scarce, we will no longer be able to run our conventional power plants at all. But the larger our supply of renewable energy, the greater the basis will be for further growth without fossil energy.

In the beginning, we will not even have to switch over to bioenergy completely, for it can be mixed with fossil sources. For instance, coal-fired plants can be operated partially with such biomass as wood or straw. Germany's Institute of Applied Ecology estimated in its 2004 study[19] that up to 10 percent biomass could be used in coal plants without major technical adjustments. Mainly, conveyor systems and related infrastructure elements might have to be adapted. In England, power plant operators are planning to start burning olive stones and palm seeds in their coal plants. Pits, stones, and fruit waste already are used to fire some power plants in southern Europe.

Some power plants have completely switched to biomass. For instance, there is a 36- megawatt plant in Ely, England, that burns straw. Here we begin to see that the energy balance does not tell the whole story, for straw has an energy balance close to 1, meaning that you do not get more energy out of it than you put in. Would it not make more sense to import palm oil or some other form of biomass that has a much better energy balance? Or maybe it would be better yet not to waste energy transporting biomass but simply to use it where it is grown?

WHICH PLANT HAS THE BEST ENERGY BALANCE?

It's true that it does not make sense to grow straw for energy. But no one is thinking of doing that, for straw is an agricultural waste product that already is available in large amounts — in many regions, more than is needed to feed animals. It can either be plowed back into the ground to return some of the nutrients to the soil or used to generate electricity. To prevent the soil from losing these nutrients, the ash can be taken from the power plant and put back into the soil.

Here we see why the figures given for the energy return on investment also do not tell the whole story. In addition, it is almost impossible to give a reliable figure for EROI because it is not always clear how much energy is needed for transport — across what distance, by what means? Since the parameters cannot even be determined, no wonder it is so hard to find reliable estimates. Nonetheless, it is a good idea to compare estimates of EROI to determine, for instance, whether transporting biomass across a certain distance is justifiable. The IEA has created Task 40 FairBioTrade to study when trading each kind of biomass makes sense.[20]

Since, as we saw above, ethanol from wood has a better energy balance than ethanol from corn, why should we grow corn as an energy plant at all? We can go even further and ask why we don't concentrate on the fastest-growing types of trees, such as willows or poplars. But then again, palm trees and *Miscanthus*[21] plants have the best energy balance of all. According to some estimates,

Miscanthus can produce up to 25-30 tons of dry mass per hectare every year (over 10 tons per acre), which is equivalent to 14,000 liters of heating oil.[22] More conservative estimates put the harvest at a maximum of just below 12,000 liters, with mean values closer to 10,000.[23] But even then, *Miscanthus* would produce twice the energy of willows or poplars. So why don't we switch over to *Miscanthus* plantations completely instead of fast-growing trees? According to one US study, *Miscanthus* has an energy balance of 3,430 percent (34.3) — compared to 121 percent (1.21) for ethanol from corn.[24] Why are Europeans planting so much rapeseed as a substitute for diesel if *Miscanthus* is theoretically so much better?

In practice, *Miscanthus* has a number of disadvantages. Hermann Hansen, a bioenergy consultant at Germany's Regenerative Resources Agency,[25] explained in interview with me that *Miscanthus* takes several years of growth before it can be harvested at all, and peak yield comes several years after that. Hansen estimated that a maximum of around 20 tons could be harvested per hectare in the best years. In contrast, a field of corn or rapeseed produces its highest yield every year, with natural fluctuations.

A plantation of *Miscanthus* thus is more like a forest: it requires more planning and is a long-term investment. But, unlike a forest, harvesting is a problem because the plant is harvested in the spring when the soil is still quite wet. At such times, machines tend to compact the soil, crushing the root systems. While soil compaction is a general problem in agriculture, it is not as severe with plants such as corn and rapeseed, which are replanted every year anyway after the soil has been completely plowed.

So we see that some plants simply are easier to manage. Palm oil might have a better energy balance than ethanol or rapeseed oil, but palm trees cannot be grown easily in Germany or Canada, and transporting the oil from the south begins to cut into the initially positive energy balance. In addition, once you start importing your energy, you become dependent. Rapeseed is planted in Germany in fields and meadows not necessarily used for agricultural purposes. And when the whole plant is used, not just the seeds, the energy balance can be increased.

If we are going to use more biomass to meet our energy needs, to ensure sustainability we will have to use a wide variety of plants. Waste products from agriculture also should not be overlooked.

SPECTACULAR BREAKTHROUGHS?

Three recent developments deserve mentioning. At this point, each of them seems to be as speculative as spectacular.

The first development concerns algae grown in salt ponds in the desert. According to the National Renewable Energy Laboratory (NREL),[26] an algae farm covering just 10 percent (38,000 square kilometers) of the entire desert in the US would produce so much biodiesel that the US would not need to import any oil for fuel. (Heating oil was not taken into consideration.) If this method proves to be effective and does not have great environmental impact (such as salt accumulation at the surface from evaporation), the area of land that could be used to produce biomass would increase considerably. The researchers are confident that all of the potential environmental problems can be solved.

The second development enables us to use entire plants as energy carriers. In the case of biodiesel, only the rapeseed is used, with the stalk of the plant still being a waste product. The Choren Company of Germany[27] has developed a way of not only using entire plants but also mixing different types of plants. This would solve the problem of monocultures and allow us to simply go out and mow a meadow when we want to harvest biomass. The mixture of plants would benefit local biodiversity.

The third development has been a hot item in the US press since *Discover* magazine reported on it in 2003.28 In 2003, Changing World Technologies made headlines in the United States and abroad with the announcement that it would be able to make oil out of just about anything. The company had been running a plant that processed seven tons of turkey offal a day into oil at a cost of around $15 per barrel. After a larger plant that would process 238 tons of turkey offal a day did not go into operation on time (because, the company said, of construction errors, not technology problems), skeptics began to wonder whether this was another fly-by-night operation. I therefore interviewed Brian Appel, CEO of the company, in December 2004 (see sidebar).

FINALLY A BREAKTHROUGH FOR OIL? A CONVERSATION WITH BRIAN APPEL OF CHANGING WORLD TECHNOLOGIES

Mr. Appel, your company is drawing great interest both in the US and elsewhere for its promise to turn waste products into biodiesel at competitive prices. If I understood your company's spokesperson correctly, you just returned from Canada yesterday, where you met with US President George Bush and Canada's Prime Minister Paul Martin.

I was invited by the prime minister.

So it seems that your company is the real McCoy and not just another fly-by-night operation that is going to offer us free energy.

We have put over $90 million into this company, and the reason other technologies are considered fly-by-night operations is that they do everything at the lab scale and are not able to take it to the next level. It takes more than just researchers; you need to look at everything from logistics to financing.

We hooked up with a big food company that was interested in using all the waste from the food chain without putting it back into animal feed. If you want to make the chain more sustainable, then you need to do things like this. When we moved from our seven-ton plant in Philadelphia to the 250-ton plant in Carthage, Missouri, we had to redesign everything.

Look at the only other success story in biofuels in the US: ethanol. Some say it's not even a success story because without the subsidies it would never work. Ethanol is an additive for gasoline, while we produce a synthetic diesel. Ethanol also has a 30-year head start. Our plant in Carthage, Missouri, is the first commercial one of its kind ever. And we are still tweaking certain parts of the design to enhance performance. I'm sure that the design of the next few plants to be built will be slightly different.

Where will these next plants be? Will they also process turkey offal?

More than likely they will process beef. The next plant will probably be in the British Isles. Remember that the British Isles received much of the blame for spreading BSE. There are now much tighter restrictions on the input side of the food chain there. Europe now protects its food chains, so we will get paid to "dispose of" the remains of cattle. In the US, we would not be paid because farmers can still take unused parts to a renderer, who will put them back into animal feed.

And then there is the output side. As you know, there is an EU directive stating that more biofuel has to be produced. In the US, the subsidies are basically for soybean and corn.

In addition, we are also able to sell a co-product as fertilizer because the United States is starting to promote organic farming. In Europe, we wouldn't get as much for this fertilizer because almost everything you do over there is organic.

Oh, we don't have that much organic in Europe.

Compared to what we do in the States, European farming is organic. Just about the whole rest of the world is farming normally compared to what we are doing. So here I'm getting a premium because there's a movement over here to buy organic. If I go to Europe and sell this fertilizer, I have to drop my expectations to the level of normal fertilizer.

Granted, without the proper management — logistics, financing, etc. — your company would not be successful. But I think most people are interested in seeing that the technology behind it all really works. When I first heard about what you were doing a few years ago, I also rolled my eyes when I read that you wanted to speed up the process of creating oil down to 15 or 20 minutes.

It takes about 15 or 20 minutes to run the process in the main reactor. But you are flattering me. We don't think our processes are that complicated.

So why didn't anyone think of this before?

We had such an abundance of light crude oil. You used to be able to stick a straw in the ground in Texas — and you still can in Saudi Arabia — and light crude oil just comes bubbling out. But a lot of the light stuff has been used up, so we're dealing with more heavy oil now.

Second, we have now had 150 years to see what the impact of the use of all of this fossil oil is going to be. And since the sixties and seventies, there has been a growing environmental movement. In the US, Rachel Carson wrote *Silent Spring*, you had the beginnings of the Green movement in Europe, etc. So after the initial denial, we have begun to think about how to become more sustainable in the past 40 or 50 years and look for solutions for waste.

So people are just now looking for technologies like yours?

Sure, but look back at some of the inventions that were made 100 years ago, especially in Germany. There was some amazing stuff. Ahead of its time.

One prime example being the fuel cell, whose basic design was developed all the way back in 1838.

Right. But the time was not right. People have been using pyrolysis for some time, but that produces a very nasty by-product, and the oil companies complain about the quality of the oil produced by pyrolysis and won't buy it. And electricity companies won't buy it because of the pollutants.

So we decided to do things backward and start with the requirements. What does it take to meet the specifications for engines? Then, we basically added refinement steps after our initial stage. Refiners do the same thing: they take crude, desalt it, separate the light oil out, etc. So you can't do this in one or two steps.

You're talking about motor engines, and most people think about cars and trucks when they hear that. But your company sells most of its oil to a power company. Is there some difference between the engine that produces electricity in that plant and the basic diesel motor in a car? From what you just said, it sounds to me like you could produce for almost any specification.

Power companies in the US have renewables portfolio requirements. So the utilities have to produce X amount of power from renewable sources. Over here, you get a 1.75-cent tax credit per kilowatt-hour of green power. And, quite simply, the company that helped us fund the pilot project needed more electricity in its renewables portfolio.

The main reason I'm asking about why the biodiesel you produce is mostly used to generate electricity is because of a common misconception. Many people are calling for more solar power and wind power because we are running out of oil. But they are missing a crucial point: with wind power, we make electricity; with solar, electricity and heat. When oil starts to become scarce, we are going to need mostly motive power. That is why the potential of biomass, and hence of your company, is so crucial.

We are working with the Big Three. Right now, we are working with DaimlerChrysler to develop a motor fuel. But we are also working to clean up the sector of heavy fuels, which causes most of the pollution. And there is one advantage to starting with stationary motors like the ones used to generate electricity: you can easily see what the long-term effects of emissions are, what the wear is on seals, fuel line filters, etc.

Right now, we're facing a situation where engines will have to be tailored to these new biofuels. For instance, if you use biodiesel in a cold area, you might find that your fuel lines clog up because the fuel has congealed. People think that the fuel is bad, but the fuel's not bad. You just have to know how to use it. My fear is that the excitement about using biofuels might backfire. So if we use a blend in a stationary engine, we can better study what the long-term effects will be.

I have been working intensively with the Defense Energy Support Center (DESC), which is the biggest buyer of fuels in the world. That's the US military. We'd like to have the post office running on biofuels and get as many people as possible involved. And there are lots of other companies like us working on synthetic hydrocarbons that can serve as a transition to take us to the next level, beyond internal combustion engines, which is what I think a lot of people are shooting for.

And what is the next level? Fuel cells?

I don't think they're possible, personally. Right now, the main supply of hydrogen comes from oil and coal, so there's a lot of hype. Here's what we care about: the company's stated mission is to clean up this waste, produce a clean fuel, and minimize global warming because much less fuel would have to be dug up from beneath the ground. If we can do that, we will have a better quality of life and cleaner air and our way of life will be more sustainable.

Energy Sink or Energy Source?

If we want to have large-scale biomass to provide energy, we may have to change our way of life. To understand why, we need to take a look at the way we eat. As we saw in Chapter 2, the authors of *Limits to Growth* point out that more energy is invested in aquaculture than we get out of it. The same is true of agriculture: we pump so much fertilizer into our fields that they actually produce less energy than we put into them. A study conducted by the Swiss Research Institute for Organic Agriculture (FiBL)[29] found that while an organic farm may produce 20 percent less product it also makes do with 50 percent less energy than an industrial farm. Of course, energy is not the only thing at stake here. We are also running out of some of the fertilizers behind the Green Revolution. For instance, the mining of phosphate has made the island of Nauru uninhabitable.

In his book *Against the Grain: How Agriculture Has Hijacked Civilization*,[30] Richard Manning describes how Americans eat tremendous amounts of sugar and protein and end up with diabetes and obesity. In other words, US agriculture is consuming huge quantities of fertilizer, chemicals, and energy so that Americans can be unhealthy. In 1940, a calorie of fossil energy produced 2.3 calories of food. Manning estimates that up to ten calories of fossil energy are in every calorie of food we eat in the US today.

Manning sheds more light on the concept of industrial food production as an energy sink. Unlike many environmentalists, he believes that we should not stop eating meat but rather switch to eating game. Like wild fish, such animals consume food that humans generally do not. In other words, the energy balance is negative only for animals raised on farms.

If we do not use so much grain to raise animals for food, we will have a lot of land left over for energy plants. We also would be healthier because the environment would be cleaner and because we would once again be eating the food for which our bodies are designed.

Chapter 5

Coal or Climate?

A simple calculation shows that the temperature in the arctic regions would rise about 8° to 9°C if the carbonic acid increased to 2.5 or 3 times its present value.

—The slightly inaccurate equation of Svante Arrhenius, winner of the Nobel Prize in Chemistry in 1903, in his On the Influence of Carbonic Acid in the Air upon the Temperature of the Ground, *published in 1897, when the concentration of CO_2 in the atmosphere was almost a third lower than today*

The History of Coal Power

In recent years, there has been much talk of "clean coal" but for centuries coal was considered a filthy form of energy. That did not stop people from using it. Indeed, without coal, industrialization would not have been possible. Coal also paved the way for the rise of oil. Thanks to coal, we no longer live in societies of subsistence agriculture. But now coal is a threat to our future. Where do we go from here?

Nowadays, when we speak of coal we immediately think of the fossil energy carrier. But in the Middle Ages coal was mainly associated with what we now call charcoal, which is basically wood that has had its water content baked out of it to make it lighter (easier to transport) and to produce higher temperatures. It wasn't until the 19th century that fossil coal replaced charcoal on a large scale. By that time, wood had become very scarce in most parts of Europe. If the British had not found a way of mining coal on a large scale (with the invention of the steam-powered pump to drain coal mines), Europe might be without forests today.

In England, you can still see how scarce wood was back then. Once a solid forest, England is now mostly rolling hills of pastures and farmland. The same is true of the rest of Europe. Italo Calvino's wonderful story *The Baron in the Trees*[1] concerns a nobleman who travels across southern Europe through the trees without ever touching the ground. Most modern readers probably consider this idea completely fanciful, but Calvino is referring to the time when Europe truly was covered with trees (albeit not in the Renaissance, when his story takes place). Some 2,000 years ago, when Germany's Black Forest got its name, parts of Europe were so densely covered with forest that it was dark there during the day. When I peer out over my laptop at the Black Forest today, I see a forest that is still beautiful in many parts but truly dark nowhere. It now has wide roads running through it, large ski slopes, and sections suffering from the effects of acid rain. A few decades ago, Germans coined the term *Waldsterben* for the deterioration of their forests caused by acid rain.

In England and France, the scarcity of wood began earlier than in Germany. In the 16th century, the British navy was already importing large quantities of timber from Scandinavia and Russia because England simply did not have enough large trees to build the many ships it needed. Clive Ponting explains in his highly recommended *Green History of the World*[2] that the English would not have ruled the seas without the trees that the navy brought back from the colonies. It was a vicious circle: the navy needed the trees from the colonies in order to build the ships to get there. By the 19th century, all English battleships were made of imported timber until gradually ships began to be made of steel.

Why wasn't France ever a major sea power? Partly because France had already devastated its forests in the 12th century to build gothic cathedrals. Before the keystone was put into the vaulted arch, the entire construction had to be held up by a wooden frame. Similarly, Greece lost most of its forests during its primacy some 2,000 years ago. Today, the country suffers from severe erosion without its natural tree cover.

England was one of the places where coal was first used for heating on a large scale (China having started much earlier). In the Middle Ages, "sea-coal" was widely used — "sea" because the coal was found in seams directly on the seashore near Newcastle and because this coal was transported by sea. In 1257, Queen Eleanor left Nottingham because she could not bear the polluted air any longer. In 1306, King Edward I banned coal from London altogether. The ban did no good; people preferred coughing to freezing. As Barbara Freese explains in *Coal: A Human History*,[3] there is documentation since 1600 that garden plants in London suffered from the polluted air. In 1661, in his manifesto against coal called *Fumifugium*,[4] John Evelyn called for a ban on "this hellish sea-coal." Apparently he was not aware that a ban had been in effect in London for some 350 years; it was just not being enforced. By 1700, British coal production was already several times greater than that of the entire rest of the world.

In the 19th century, pollution from coal began affecting Germany. In 1876, Robert Hasenclever wrote what is probably the first scientific study of air pollution, *Über die Beschädigung der Vegetation durch saure Gase.*[5] He spoke of "acid gases," which had been known in common German parlance as *Hüttenrauch* or "cottage smoke." In 1905, the term "smog" was coined in England from the two words "smoke" and "fog." The smoke from coal stoves regularly darkened the days in England and North America in the first half of the 20th century. Smoke control regulations eventually provided for cleaner air, but not until the city of London had four days of continuous darkness in December 1952 when some Londoners reported they could not even see their own feet. In the late 20th century, the term "acid rain" was coined, initially in reference to the forests of Scandinavia that were suffering from pollution from the coal plants in Britain.

The plethora of terms — acid gases, smog, acid rain — demonstrates that there is no continuity. It seems that people do not want to see the connection between the smoke that darkens the sky and the warmth from the coal. As Barbara Freese writes, people long believed that the smoke from their stoves — from either coal or wood — cleaned the air much as it kept smoked meat fresh.

Can Europeans and North Americans today even imagine what it is like to have to choose between keeping warm and breathing clean air? That is exactly the choice that millions of people still face every day, as Mark Hertsgaard writes in his *Earth Odyssey* of 1999.[6] He found the air in Beijing to be the worst he had ever breathed and tells a story of a Western diplomat who continued jogging during his stint in Beijing. When the man returned home, his doctor told him to stop smoking; the diplomat was a non-smoker. Then Hertsgaard visited Taiyuan and Chongqing and no longer knew where the air was the worst. In the latter city, people told him there are days when you can't even see your own hands.

According to the World Health Organization, the volume of airborne particles should not exceed 40 to 60 milligrams per cubic meter. In 2002, Chinese environmental authorities reported that two-thirds of the cities studied exceeded these maximum values. In some cities, the level never fell below 300 milligrams, with peak values of up to 900 milligrams.[7] According to a report in the *Financial Times,*[8] 50,000 Chinese die each year from the effects of emissions from coal power plants and stoves in the country's 11 largest cities; by 2010, this figure could increase to 380,000 — and by 2020 even to 550,000. The scenario is not improbable, for China has tremendous coal reserves and currently is the world's largest coal producer.

Clean Coal?

Why don't the Chinese just switch over to our cleaner coal technology? The answer is quite simple: it would make their coal power more expensive. Indeed, we can ask ourselves in the West a similar question, for our coal plants may be cleaner than those in developing countries but our current plants are not nearly as clean as technology would allow today.

The US Department of Energy plans to have clean coal plants as the standard by 2020 without making power significantly more expensive. The FutureGen project aims to put such plants on the grid in the next ten years.[9] In the presidential elections of 2004, George W. Bush increased his support for clean coal from $1.6 billion to $3.7 billion, while candidate John Kerry went so far as to promise $10 billion.

The coal industry has been profitable since the 19th century, but it currently is calling for state subsidies for programs that finally will make it relatively clean. At the same time, for decades the coal industry has been demanding that its current plants be "grandfathered" — made exempt from new environmental standards. According to a study in 2000 conducted by the US coal industry,[10] it would have cost some $65 billion to equip all the country's coal plants to meet the latest environmental standards. The cost for each plant would have amounted to hundreds of millions of dollars.

Figure 5.1:
The German coal industry supports renewables. If this book could be summed up in one sentence, it would be: Let's use our remaining finite resources to set up a renewable energy supply as quickly as possible. The German coal industry sees it the same way. This ad reads: "You need time. We will give it to you." Fossil fuels should not be stamped as the bad guys, but used as a platform for a renewable future. (Source: Deutsche Steinkohle AG)

The latest technology is called integrated gasification combined cycle (IGCC, discussed later in this chapter). Critics of this technology point out that such projects could make coal power so expensive that we might as well switch to renewables to begin with. Today, wind power often costs only five cents per kilowatt-hour in good locations, with biomass sometimes costing less than eight cents. Tidal power and wave power also may be this affordable once the technology matures.

Coal Power Today

Coal is the number one source of electricity worldwide. Unfortunately, in the past few decades it has become clear that carbon dioxide — one of the main emissions from coal-fired power plants but previously not monitored as a pollutant — poses a great danger. Carbon dioxide is a greenhouse gas that contributes significantly to global warming. The industry has been focusing on scrubbing carbon monoxide, sulfur dioxide, and nitrogen oxides out of coal emissions, but until recently little attention was paid to carbon dioxide. Today some 70 to 90 percent of sulfur dioxide can be removed from a coal plant's flue gas. Carbon sequestering is widely held to be the solution for the carbon dioxide problem. In this process, the carbon dioxide that a plant produces is separated, possibly liquefied, and pumped into storage areas underground such as gas and oil fields.

Ideally, the carbon dioxide would be injected back into the subterranean field it came from. What sounds like a brilliant idea actually works, as the project by Norway's state-run oil firm Statoil shows.[11] Since 1996, Statoil has been separating carbon dioxide from the natural gas and oil fields under the North Sea and pumping it back into the reservoirs. This method even has a nice side effect: the pressure in the subterranean field is kept high, making it easier to pump the natural gas. But even in this case, the project only pays for itself because Norway was one of the first countries to impose a substantial tax on carbon dioxide emissions — in 1991, long before the Kyoto Protocol.

Proponents of this approach rarely mention the obvious contradiction: the process of injecting CO_2 back into the oil and gas fields to keep the pressure up adds to CO_2 emissions. The major sequestration project at the Weyburn oil field in southern Saskatchewan, Canada, is an excellent example of this contradiction.

WHAT'S THE CATCH?

While sequestration may work in the oil and gas industry, it may not in the coal industry because CO_2 emissions from coal mainly occur at the power plants, not where the carbon dioxide might be stored. In addition, it is not at all clear how much carbon dioxide we would be able to store underground and at what price. According to a report in the British daily *The Guardian,*[12] Great Britain could store its entire carbon dioxide emissions for at least ten years under the North Sea. Greenpeace has estimated that Germany could store only around seven years' emissions within its own borders.[13] However, we have enough coal to last for several centuries.

Globally, the situation is not much better. The Intergovernmental Panel on Climate Change (IPCC)[14] has estimated that the world could sequester some 400 billion tons of carbon. If we divide that by the approximately seven billion tons of carbon we emit every year, we apparently can sequester our emissions for only around 57 years. As author Lindsey Grant put it for CO_2: "What do you do with some 16 billion tons of CO_2 a year? (For a sense of the scale, consider that that is eight times as much tonnage as all the world's annual grain production, and it is vastly larger because it is a gas.)"[15]

And it is not going to be cheap. Estimates are that sequestering will cost between $80 and $420 per ton of carbon dioxide. In the Kyoto Protocol, a ton of carbon dioxide will only cost around six dollars, and the Norwegian tax that made that country's sequestering project possible also "only" charges around $50 per ton of CO_2. In the oil and gas industry, that difference can be compensated for by greater pump production, which again ironically partly defeats the purpose of sequestering by releasing even more carbon dioxide into the atmosphere. But when it comes to coal, sequestering will only increase costs. Estimates are that at $100 for a ton of carbon dioxide, the price of a kilowatt-hour from a coal plant would double.

And what will we do once our capacities have been exhausted in the mid 21st century? No problem, say the researchers in the EU project JOULE II,[16] which came to much different conclusions in the 1990s. They estimated the storage capacity within the EU and Norway alone at 800 billion tons of carbon dioxide. The main difference in their calculation was that they included aquifers in addition to exhausted oil and gas fields. The project also believed that the oceans could store around one trillion tons of CO_2 without any nasty consequences. But in the summer of 2004, the US National Oceanic and Atmospheric Association[17] announced that by 2100 the oceans could have the lowest pH value in the past five million years because of the absorption of carbon dioxide that is already underway. The low pH value may soon prevent shellfish from forming shells.

COAL POWER PRODUCES MORE CARBON DIOXIDE THAN POWER

In an interview with me, Jeffrey Michel, author of the 2005 study *Status and Impacts of the German Lignite Industry,*[18] explained another aspect of the dilemma this way: "A train pulls up to a power plant. This train is supposed to take away the carbon dioxide that the coal plant produces when the coal has been unloaded. Normally, though, the train that takes the carbon dioxide away from the plant would have to be longer than the train that brings the coal because one ton of coal produces slightly more than one ton of carbon dioxide on the average because the carbon in the coal combines with oxygen in the atmosphere. Who is going to pay for that?"[19]

Michel's work has brought attention to the eastern German village of Heuersdorf, where he has been living for about ten years. *The New York Times*[20] reported on the village, which is to be sacrificed so that a coal mining company owned by two US holding firms can get at the coal underneath the village. Families who are willing to sell their homes to the mining company are offered a fair market price plus around 75,000 euros. Many have already moved away, but not without some major family disputes in cases where the children wanted their parents to take the money and run but the parents wanted to spend their retirement years where they had spent their lives. The offer for the extra 75,000 euros only applies to the people originally living in the homes when the offer was first made; in other words, if the parents die, their children cannot move into the house and sell it to the mining company. It is a strategy designed to clear out the village as quickly as possible. In some cases, spouses did not agree on what to do.

It is worth considering the effects of mining when we talk about whether we want to have coal power or renewables. A wind farm could never destroy an entire village. Nonetheless, detractors of wind power charge that it spoils natural landscapes. Take a look at what coal mining does to the countryside.

Figure 5.2:
The village of Horno — or rather what is left of it after it was destroyed to mine coal underneath it. Such devastation is widespread in the US, where strip mining is common. (Source: Photo courtesy of Gerard Petit)

There is no denying that wind turbines change the way a landscape looks. However, it is a matter of taste whether the countryside is temporarily destroyed in the process. In the case of coal there can be no doubt that landscapes are completely destroyed forever. It takes decades for the lunar landscapes that strip mining leaves behind to be recultivated and even then the landscapes are never restored to their original condition.

Even if we believe that wind farms completely ruin natural landscapes, we still have to admit that wind turbines can be dismantled at any time without leaving a trace. Wind turbines do not change anything forever. Not so with coal power: Jeffrey Michel estimates that 1.1 billion tons of earth and coal are excavated every year in coal mining in Germany alone — 15 times as much as the 74 million cubic meters of earth excavated when the Suez Canal was constructed.

Coal mining practices are no worse in Germany than elsewhere. At present, Germany is importing more and more hard coal because its domestic reserves are simply too expensive. As a result, in recent years German coal miners have been migrating to the United States, where coal mining is booming. Part of the reason for the success of coal power in the US is that mining companies are allowed to completely devastate the landscape without paying the environmental costs. In the US, the term "mountaintop removal" has been coined for the brutal strip mining that has been taking place in the Appalachians and elsewhere. There are grassroots movements against such devastation, which leaves giant tracts of land damaged for centuries and can also cause disasters that threaten human lives. For instance, on October 11, 2000, some 250 million gallons of coal-laden sludge came sliding down a mountain in the Appalachians for about 100 miles. Groundwater was poisoned for years by the sludge, though miraculously no one was killed.

But this common practice in the US is still far better than everyday practices in China. In the US, state-of-the-art technology is used and safety standards are relatively strict to protect the lives of miners. In China, a number of mines are operated by small and mid-size businesses, with miners doing basically everything with muscle power. Rarely do safety measures ensure that the mountain does not collapse, trapping the miners in their own graves. According to Barbara Freese, 51 coal miners died in accidents in the US in 1992, compared to around 10,000 in China in 1991.

The number of coal miners who die from the effects of coal mining really is much higher. While the number of miners who die from accidents in the US has dropped to 30 per year, around 1,500 per year still die from inhaling coal dust. One out of every 20 miners suffers from respiratory problems. In addition, coal plant emissions have been the cause of 554,000 cases of asthma, 16,200 cases of severe bronchitis, and 38,200 heart attacks in the US alone.

According to recent studies, the health costs from these emissions amount to $160 billion.

We should not forget the environmental impacts of coal plant emissions. While a wind turbine kills a few birds every year (see Chapter 9), coal power appears to endanger entire species. For instance, populations were decimated on the Shetland Islands in the summer of 2004 when the 24,000 arctic terns and 17,000 kittiwakes that build nests there produced no eggs, while the 350,000 guillemots produced only a few young. Researchers discovered that the water in the North Sea has become 2 degrees Celsius warmer in the last 20 years and many species on which these birds feed have moved into colder waters.[21]

The environmental impact of coal power extends beyond plants and animals to directly affect humans. The US Environmental Protection Agency (EPA) estimates that some 24,000 Americans die from the effects of emissions from coal plants every year.[22] Of these, 22,000 could be saved if the strictest "clean coal" technologies were implemented. According to the EPA, 63 percent of the sulfur dioxide emissions in the US, which cause acid rain, and 40 percent of mercury emissions come from coal plants. The extent to which coal plants impact the environment became measurable by accident in August 2003 during a blackout in the US. Coal plants in the northeast were shut down and within 24 hours concentrations of sulfur dioxide in the air in Pennsylvania dropped by 90 percent.

While the coal industry likes to complain about the costs of stricter standards for emissions, for society as a whole the avoided health costs greatly outweigh the extra costs of cleaner coal. The EPA claims that by 2020 the controversial Clear Skies campaign will lead to around $18 in health savings for each dollar invested.

Coal as a Bridge to a Renewable Future

As we have seen, coal has long been hated for the environmental damage it causes but loved even more for the heat it gives off. If there is no alternative to coal today, will there be one tomorrow — and if so at what price?

James Lovelock, the principal creator of the concept of the Earth as a single living organism, made headlines in 2004 when he called for more nuclear power to prevent impending doom from global warming (see Chapter 6).[23] He set off a debate about the pros and cons of nuclear power, but almost everyone overlooked the reason he had called for more nuclear power: the Earth has to be saved from coal power.

What does climate change have to do with coal plants? No one can "prove" that emissions from coal plants are chiefly responsible for global warming. Skeptics in the US like to point out that there have always been temperature fluctuations and that we do not fully understand the effect of solar flares, but it would be astonishing if the increase in the share of carbon dioxide in the atmosphere during industrialization (from around 280 parts per million to 384 parts per million) had not led to the greenhouse effect.

In the meantime, the debate about whether climate change is happening and why, is no longer being conducted outside the US, and continues inside the US only among those with vested interests in fossil energy. We have a choice: we can kill a few birds a year with each wind turbine or entire species with coal power.

But we must also remain realistic. If coal is so bad, why not just get rid of it altogether? After all, it contributes to global warming, creates air pollution, and destroys landscapes during mining. However, coal power is cheap as long as we do not calculate the external costs. Perhaps more important, unlike oil and gas, which will become scarce in the next few decades, our coal reserves will last for a few more centuries. But the reason I insist on remaining realistic is that people probably still will be burning coal even after the worst nightmares of global warming have come to pass. At any rate, even if we did away with coal power, Lovelock himself admits that it would take decades before the concentration of carbon dioxide in the atmosphere would be reduced.

We therefore have to start thinking about how we can use coal more efficiently. As a non-renewable resource it should be treated with care. Like oil and gas, it is a one-time gift — we will never get it again — so we should not be burning it as though there were no tomorrow. If we can extend the range of our coal reserves from 200 to 700 years, our descendants will thank us. Too bad it is too late to do this with oil.

We could start by including the external costs of coal power in the prices we charge to consumers. This would help make reducing emissions from coal plants, especially carbon dioxide emissions, affordable. It might also mean that coal power would become as expensive as many renewables. This is the good news: renewables would no longer need subsidies, and power would generally be expensive enough that everyone would start thinking about efficiency to keep costs stable.

A modern coal plant has an efficiency of around 35 percent. In other words, just over a third of the energy the coal contains actually is converted into electricity. The rest is converted into heat that generally is not used ("waste heat"). More energy is lost than gained. The potential of efficiency is greater than that of oil, gas, and coal combined. The question is just how much of this waste heat can be used in practice. Coal plants have one major drawback over cogeneration units, where waste heat is used for domestic heating or industrial purposes. Coal plants often are very large — hundreds of megawatts — and rarely are close to large settlements. Power plants fired

KEY FIGURES FOR AN AVERAGE COAL PLANT

A modern coal plant with a capacity of 500 megawatts consumes annually:

- 1.43 million tons of coal
- 8.3 billion liters of cooling water

Only 3 percent of the water evaporates. The other 97 percent exits the power plant and generally is fed back to the body of water it came from at least 10 degrees Celsius hotter than when it entered, which poses a risk to plants and animals. Coal plants therefore cannot be used easily in areas with little water. One of the main problems with the expansion of coal power in the US is that the areas that need more electricity, such as Las Vegas, simply do not have the water resources to cool the plants. In the record hot, dry summer of 2003 in Europe, when the city of London recorded a temperature of 100 degrees Fahrenheit for the first time in its history, France and Germany had trouble keeping their nuclear and coal plants in operation for lack of cooling water. As a result, the admissible temperature in some rivers and lakes was exceeded.

The plant emits annually:

- 3.7 million tons of carbon dioxide — more than 2.5 tons for every ton of coal (causing global warming)
- 10,000 tons of sulfur dioxide (acid rain)
- 10,200 tons of nitrogen oxides (smog)
- 720 tons of CO (indirectly increasing global warming)

A forest used as a carbon sink would have to cover 2,000 square kilometers to store all the carbon dioxide that such a coal plant produces over the course of its service life. The carbon dioxide absorbed by these trees would eventually be released into the atmosphere, either when the wood decomposes after the trees die, which could take centuries, or as quickly as overnight in the case of a forest fire. In January 2005, the Pew Center for Climate Change summed up the inadequacy of forest carbon sinks when it wrote that removing one-fifth of annual US carbon emissions would require an area the size of Texas. (Sources: Union of Concerned Scientists and JOULE II) with brown coal (lignite) generally are placed close to the mines because the water content in brown coal is so high that transport costs would be excessive. When such gigantic central plants are too far from large populations, waste heat cannot be easily used to heat households.

When it is not economically feasible to use waste heat from coal plants for heating, there are other ways of increasing the efficiency of coal power. Recently, there has been much talk of "syngas" (synthesis gas): gasified coal more or less as it was used 100 years ago when coal was king and natural gas was still considered a waste product (see Chapter 7). In gas-and-steam turbines, this syngas drives one turbine while the waste heat drives another. Such a facility is being used at a coal plant in North Dakota, where the syngas is made from brown coal.[24] Europe also has such plants, including the DEMKOLEC project in the Netherlands, which in 1994 produced a power plant with 253 megawatts of electrical output (MWe) that even purifies the wastewater created during the gasification of coal.[25]

The principle of coal gasification is easy to understand. Under high pressure, water and oxygen are added to the coal, which is cooked into its useful components of carbon monoxide and hydrogen. This mixture of carbon monoxide and hydrogen is the syngas. Waste products such as carbon dioxide and sulfur can be separated out and disposed of during this process. As this separation takes place anyway during gasification, carbon sequestration is facilitated. In addition, at least 90 percent of the mercury in the coal is removed in this process.

Another method being tested, for instance at the University of Stuttgart in Germany,[26] uses water and calcium oxide (CaO, also known as lime, quicklime, or burnt lime). The CaO, which already is used in scrubbers in conventional coal plants, binds the carbon dioxide and turns into limestone. Only hydrogen is left over. The limestone can then be combusted in a second reactor to release the carbon dioxide so it can be sequestered or sold to industry. The leftover burnt lime can be recycled again and again to bind carbon dioxide. The jury is still out on which of these two technologies — or perhaps even another one — will prove the more successful.

Whatever the method, coal gasification offers a number of advantages:

1. The syngas can be used in conventional cogeneration turbines, whose efficiencies range from around 60 to 70 percent, almost twice as high as the roughly 35 to 40 percent efficiencies of modern central coal plants. While part of the energy that the coal contains is lost during gasification, the great increase in efficiencies at the plants more than compensates for these conversion losses.

2. As gasification allows coal power to be used in gas turbines, coal power becomes more useful in covering fluctuating loads. At present, coal plants, like nuclear power plants, are gigantic units that cover the base load and cannot quickly adapt to fluctuations in demand. Since gas turbines can be switched on

and off very quickly and efficiently, gasification will mean that coal power will be better able to compensate for fluctuations not only in demand but also in the supply of an energy system dominated by intermittent renewables.

3. The components of coal that cannot be used to generate electricity are separated out for disposal or for other applications. For instance, the sulfur from the IGCC Wabash River plant is processed into fertilizer, essentially turning a waste product into something that creates revenue instead of expenses.[27] Indeed, even CO can be used in stationary solid oxide fuel cells, which run on natural gas, gasified coal, etc., which they internally reform into hydrogen and carbon monoxide. (See Chapter 11 for a discussion of fuel cells.)

4. Coal also could be used to create hydrogen, making coal power useful for mobile applications powered by PEM fuel cells.

In a few decades, we may have even more gas turbines than coal plants. At the moment, there is much talk in the US of building new coal plants because the country's decision in the 1990s to focus on natural gas has made it vulnerable to the recent price increases. In addition, the United States does not have great natural gas reserves, making it dependent on imports, mostly from Canada. Perhaps we will find that we want to use gas turbines for their efficiency and flexible operation, but countries such as Canada with great natural gas reserves may fire them with natural gas while the United States may fire them with gasified coal. If so, there will still be a place for our centralized coal plants, which also can cover the base load with a blend of biomass and coal, a process already being used from Scotland to southern Europe.

The coal power of the future is not going to be as cheap as today's but it also will not produce nearly as many waste products. The higher price will mean that, since less energy will be consumed as people try to keep costs down, future generations will still have this finite resource. In addition, people will begin to switch over to renewables, which will no longer be seen as uncompetitively expensive and will remain available in quantities we can never exhaust. If we use the solar energy we have today, we will not need as much "solar energy from yesterday." We need to concentrate on using coal power to set up an infrastructure of renewable energy — a strategy that, at least in Germany, not only is the slogan of environmentalists but was also part of a recent press release from the coal industry itself: "Germany's hard coal can serve as a bridge between today's energy supply and the future, which will consist of wind, water, and solar." [28]

As we see, Germans across the energy spectrum are all pulling in the same direction. The only thing that remains unclear is which of these technologies will be ready for the market first. In North America, there are energy experts who believe that coal power in combination with the sequestration of carbon dioxide as described above will be a crucial part of the upcoming "hydrogen economy" in addition to wind power. As Amory Lovins put it, "Wind, and perhaps carbon-sequestered coal, will beat natural gas for making hydrogen, which will emerge as the dominant energy carrier." [29]

If we remain realistic, we must hope he is right because the way we are burning coal today is destroying our climate, and the countries that have great coal reserves — mainly the US, China, and South Africa — are as unlikely to resist the temptation to burn these riches as the English were in the Middle Ages.

HYDROGEN ECONOMY?

The hydrogen economy is a vision of the future that is not clearly defined. Most people generally understand it to mean that hydrogen will be available at filling stations much as gasoline is today, that there will be hydrogen pipelines, and that "excess" energy from renewable sources will be stored in the form of hydrogen. This hydrogen economy is said to be environmentally friendly, with no emissions from the tailpipe except pure water.

Critics point out that the problem of emissions will not be solved by switching to hydrogen. Rather, if we get our hydrogen from fossil sources such as coal we will simply be shifting emissions from the tailpipe to the power plant. Only in cases where local emissions must be kept to a minimum does such a switch provide true benefits. Critics also say that producing hydrogen by means of electrolysis is so inefficient that we would probably have lower emissions if we simply used the electricity directly without converting it into hydrogen.

The alleged benefit of hydrogen for storing excess electricity addresses a problem we do not have. Only in parts of northern Germany and Denmark, where the wind occasionally blows so hard at times of low energy consumption (such as at night) that the grid is almost overloaded, and in small grids in remote areas is there any need to store electricity at all. In the rest of the world, we do not by any stretch of the imagination have so much renewable power that we do not know what to do with it. Today, if we are producing more power than is being consumed, we simply run down all our gas turbines, followed by our nuclear and coal plants.

In the near future, if we have more renewable energy we will simply be cutting into the output of our conventional fossil and nuclear plants rather than creating excess electricity that has to be stored. The rare exceptions do, however, make headlines. For instance, Iceland is interested in the development of hydrogen because the country truly has excess energy from its plentiful geothermal resources that far exceeds what a population of only a few hundred thousand people can consume. Iceland hopes to be able to export this energy as hydrogen.

The Climate

In August 2002, the center of the city of Dresden was flooded after 6.2 inches of rain fell in one night — far more than twice the old record for a 24-hour period. "Do we have to wait for further proof of climate change?" asked Harald Schützeichel, CEO at SAG, Germany's largest solar power plant planning company, just after the flood.[30] His statement reflected the sentiment of almost all Germans, and the sentiment was only reinforced a year later when the country experienced the record hot, dry summer of 2003. Have I forgotten Hurricane Lothar, which still had the strength to devastate southwest Germany after it had denuded entire mountains in France on December 26, 1999? Does December sound like hurricane season? Actually, December is as good a time as any in Europe because Europe has no hurricane season.

Remember the hurricane that appeared off the coast of Brazil in 2003 — the first one ever recorded south of the equator? And what about the flood in China in August 2002 or the one that covered Mozambique in 2000? Are the glaciers not melting worldwide? And was the record hurricane season in the Gulf of Mexico in 2005 not proof that global warming is causing warmer oceans to produce more intense storms?

When we ask if any of these catastrophes proves that global warming is happening, the answer has to be "no" because we must not confuse the weather with the climate. Critics of the concept of global warming — who are extremely few and far between but do receive equal coverage in the "balanced" US media — like to argue that no one has proved that mankind is responsible for global warming. As George W. Bush once put it, no "sound science" has proved any such link. He was right. Science does not ever prove anything; it merely disproves theories and comes up with new ones that are considered workable until disproved themselves. Newton's ideas about physics were completely wrong but they served us fairly well until physicists such as Einstein began to show how some things could not be explained by Newton's laws. The questions that Einstein and others were not able to answer exhaustively are at the heart of physics research today. Quite possibly, the theory of relativity may one day be disproved.

Science is not about having right answers. It is about coming up with new results that even your critics can reproduce reliably. It is about keeping an open mind and admitting you are wrong when you are finally proved to be. In US politics, such healthy attitudes are belittled as "waffling," but refusing to admit you are wrong does not mean you are right.

It is impossible to run experiments on the climate of the Earth beyond the computer models used today. We simply do not have an experimental planet at our disposal. So even if the temperature has increased by several degrees by 2100 and major catastrophes have resulted, we will still not be able to say for certain that the burning of fossil energy during the Industrial Age was the main culprit.

What can we say for sure? There can be no doubt that the greenhouse effect exists. If it didn't, the average temperature on the Earth would be around -18 degrees Celsius instead of just below 15 degrees Celsius. It is also clear that the concentration of the gases that cause this effect has been increasing rapidly since we began burning fossil fuels. In the past 450,000 years, the concentration of carbon dioxide in the atmosphere has ranged between approximately 175 and 300 parts per million.[31] In the last 50 years, this concentration has been skyrocketing and we should reach 400 parts per million in the next decade or so. Scientists hope that we will be able to stop this trend short of 550 parts per million by 2100 — and fear that we will not.

Climate experts also agree that there will be more natural disasters in the 21st century than in the 20th century. If sea levels rise, we are not just going to have less beach. In addition to entire islands and coastal regions being inundated, a large part of that water will rain down on the continents in a deluge. Paradoxically, dry spells may also be longer and the increase in rainfall might not help because so much rain will fall in such a short period that the desiccated ground will not be able to absorb it. Floods will be followed by droughts and vice versa.

The world's largest reinsurer, Munich Re, has calculated that the damage from natural disasters has increased eightfold in the past 30 years.[32] The company predicts that, if this development continues, the damage will exceed the world's gross product by 2060. The company estimated that damage from natural disasters in 2004 amounted to $40 billion, partly because of the extraordinarily large number of hurricanes in the Caribbean. In 2003, the damage was estimated globally at $15 billion. The damage just from hurricanes Katrina, Rita, and Wilma in 2005 was several times this amount.

What's Hotter: The Climate or the Debate About It?

Why don't all these catastrophes prove that the climate is changing? Worldwatch Institute writes:

> With global average temperatures climbing to 14.5 degrees Celsius, 2002 was the second hottest year since record-keeping began in the late 1800s. The nine warmest years on record have occurred since 1990, and scientists expect that the temperature record set in 1998 will be broken with a new high in 2003. Scientists predict that higher global temperatures will translate into a number of extreme weather events. The number of big weather catastrophes worldwide has quadrupled since the 1960s, a trend that many attribute to rising global temperatures.[33]

The high in 2003 fell short of the record set in 1998, but only barely. Worldwatch Institute is on the mark nonetheless because it is citing figures for the planet as a whole. As the movie *The Day After Tomorrow* showed, climate change does not necessarily mean that any particular place is going to be warmer. Rather, it means the climate is going to change. This difference between warming and change is not trivial, for critics of the idea of global warming use specious examples of local weather events to disprove global warming. For example, the George C. Marshall Institute published a report at the beginning of 2004 claiming that 1,000 years ago the Earth was even warmer than it is today.[34] The "Harvard researchers" — as the authors of the report were called in the press — allegedly disapproved the claims made in the 1,000-page report produced by the IPCC in 2001. Five hundred international scientists and 300 international reviewers worked on the IPCC report, which found that the Earth had been warmer in the 1990s than at any previous point in time that can be measured today. The small group of authors from Harvard "disproved" this theory by pointing out that Greenland had been much warmer between 1000 and 1500 than it is today. In doing so, they used the weather in Greenland to talk about the global climate.

The Harvard authors are an example of a group that is not unbiased on this issue. The George C. Marshall Institute, like other critics of climate change such as the Competitive Enterprise Institute, is supported by major oil companies in the US, particularly ExxonMobil. Some of these "organizations" sound legitimate, such as the Science & Environmental Policy Project (SEPP), but the project's newsletter[35] shows how absurd the arguments are. For instance, despite all the reports that Alaska is sinking in solid tundra, SEPP points out that the website of the Alaskan Climate Research Center does not support the claims of the "liberal" *New York Times,* which stated that temperatures in Alaska had risen by seven degrees Fahrenheit. Instead, the website stated that temperatures had "only" risen by 2.7 degrees Fahrenheit since 1971. SEPP uses this discrepancy to speak of "global warming fiction."

Charles Dudley Warner's flippant remark that everyone complains about the weather but nobody does anything about it is poignant today. We can change the laws of the market but we cannot change the laws of nature. So what should we do?

On the one hand, we have the doomsday evangelists of peak oil and the "die-off" theory.[36] They claim that business as usual is in danger of destroying civilization and sending mankind back into subsistence agriculture. Critics of this camp point out that as long as records have been kept, people have been claiming that the sky is falling. According to some, such as the Danish environmental minister and author of *The Skeptical Environmentalist,*[37] Björn Lomborg, global warming is not our main problem and may even make the world a better place, such as when warmer temperatures promote greater plant growth. *Science*[38] recently reported that because of warmer temperatures the planet now has 6 percent more biomass than 20 years ago.

The problem with this theory — aside from the problem of droughts and the impossibility of predicting exactly what will happen where — is that plants on this planet have adapted to a specific temperature range. Above 35 degrees Celsius, photosynthesis gradually becomes inefficient and few plants can withstand 45 degrees Celsius at all. When temperate Europe experienced an unprecedented heat wave in 2003, vegetation was reduced by 30 percent.[39] Now imagine what will happen in areas that are hot already.

Most people come in somewhere in between these two extremes and mainly want to know what we can do at what price. The current US administration is trying to portray measures to fight global warming as bad for US business and to obliterate the connection between burning fossil fuels and global warming. In one climate study the Bush administration had all references to this connection deleted.

One thing seems certain: if we do nothing today, future generations will pay a very high price indeed. At the same time, the measures we take today to reduce emissions of greenhouse gases will probably not have a major effect during the lifetime of anyone reading these lines. Reducing such emissions thus would be almost an act of altruism were it not for the concomitant benefits of switching to a renewable system: job creation, less pollution, and greater energy independence.

But before we talk about such options in greater detail, we need to look at the role that nuclear power could play and that natural gas already is playing.

Chapter 6

Nuclear Power

> The German nuclear industry, which supplies about 30% of the nation's electricity, employs 38,000 people. The German wind industry therefore produces 10 times as many jobs per unit of installed capacity, and *more than 20 times the jobs* in terms of delivered electricity.
>
> —*US energy consultant Donald Aitken, 2005*

Reports of the Death of Nuclear Power Are Highly Exaggerated

Shortly after the Social Democrats and the Greens took office in Germany in 1998, this left-of-center coalition resolved to decommission the country's nuclear power plants after 32 years of operation rather than extend their service lives beyond the initial plan of 40 years. While this step has been publicized as the "phasing-out of nuclear power" (*Atomausstieg*), since the power companies themselves helped design this exit from nuclear power they were at least satisfied to be getting out of the costly technology in an orderly fashion without going bankrupt. (A grand coalition of the Social Democrats and the Christian Democrats was elected in 2005. There was talk of extending commissions for nuclear plants back to 40 years, but power companies were refusing to pass on the estimated savings of up to 62 billion euros to consumers. Since this extra profit could be considered an illegal subsidy according to EU law, legal disputes may prevent the extension to 40 years.)

How will Germany do without nuclear power, which currently provides almost 30 percent of its electricity supply, by 2025? Germany gets less than 1 percent of its electricity from oil-fired plants (in the US, the figure is below 4 percent), and oil is getting scarcer. Coal plants, which make up the largest share of electricity in Germany and in the US, emit not only pollutants but also carbon dioxide, which is accelerating global climate change. In contrast, nuclear power emits neither pollutants nor greenhouse gases, which is why its supporters have been selling it as an environmentally friendly source of energy. Aside from its waste, and as long as nothing goes wrong, it is.

For several years now, the nuclear industry has been promoting nuclear power globally as a clean source of energy without an alternative. The World Nuclear Association recently even included in its logo the slogan "Energy for a Sustainable Future." The Nuclear Energy Institute has been focusing its public relations work on young women now that surveys have found worldwide that this target group is least interested in supporting nuclear power. The industry has not yet won over many people. In a 2004 poll in Germany published in the magazine *stern*, 79 percent of those surveyed were against adding new nuclear power plants, while 51 percent wanted the plants currently running to be decommissioned on schedule.[1]

At least we do not have to listen so often anymore to people saying that nuclear power is cheap, much less "too cheap to meter." But is there really no alternative to nuclear power? Or, to put it differently, even if we forget about the potential dangers of nuclear waste, accidents at nuclear plants, the proliferation of fission material both for nuclear missiles and for "dirty" bombs, and terrorist attacks on plants, would anyone really want to have nuclear power if the totality of renewable sources of energy could replace it?

Fighting for Its Life

After the highly publicized accidents at Three Mile Island and Chernobyl, nuclear power quickly fell out of favor. The accidents in the Soviet Union at Mayak and Chelyabinsk, kept secret at the time, are still relatively unknown. Since 1993, no new nuclear power plant has been put into operation in the US, and none has been ordered since 1979, the year of the accident at Three Mile Island. In Europe, more nuclear power plants will be decommissioned than built in the next few decades. Worldwide, the number of nuclear power plants online has remained relatively stable over the past ten years at around 440. In the heart of Europe, the dispute about the new nuclear plant in the Czech town of Temelin, which drew protests not only in the Czech Republic but also in neighboring Austria and Germany, shows that Europeans still do not trust this technology.

ENERGY FUNDING: WHERE WOULD WE BE?

In the past few decades, government funding for fossil energy has been relatively moderate, but we must not forget that coal and oil have been economically viable for more than a century. Even nuclear fusion has received more government funding than renewables. And while fusion is not expected to be commercially available until 2050, although it has been heavily funded for decades, renewable energy is expected to make up some 20 percent of the energy supply in the EU by 2020. Where would we be today if renewable energy had been supported as much as nuclear power in the past 30 years?

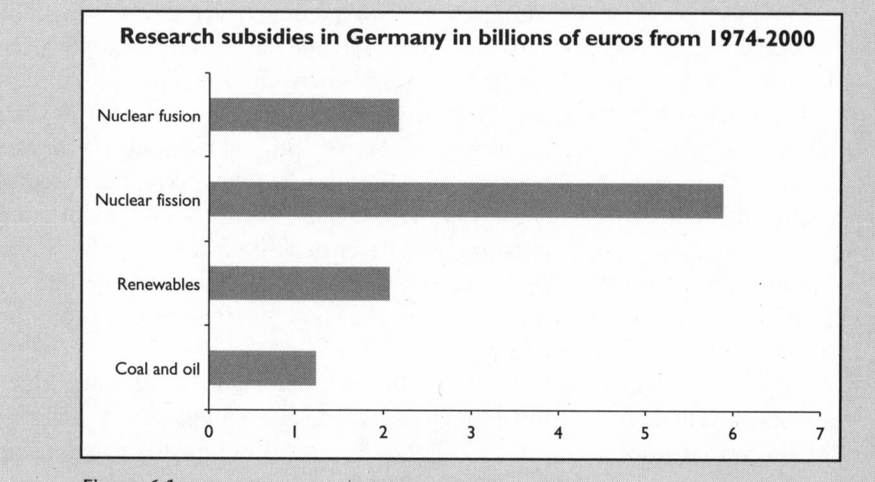

Figure 6.1:
Research funding for energy sources in Germany. (Source: *Energiehandbuch: Gewinnung, Wandlung und Nutzung von Energie*, ISBN: 3-540-41259-X)

If the 2004 study *Renewable Energy — Market and Policy Trends in IEA Countries* by the International Energy Agency[a] is any indication, we have not learned our lesson. The funding for research in renewables dropped by two-thirds (adjusted for inflation) from 1980 to 2001. Renewables still receive less financial support than fossil fuels, according to a study by the European Environmental Agency based on figures for EU-15.[b] In 2001, some 13 billion euros were invested in the coal industry, 8.7 billion euros in oil and gas production, and 2.2 billion euros in nuclear power — compared to only 5.3 billion euros for all kinds of renewable energy.

In addition, when we speak of development aid in an energy project, we often mean funds that flow back to Western companies. The Sustainable Energy and Economy Network (SEEN)[c] estimates that some 80 percent of World Bank oil projects are devoted to exports back to the West, while the Institute for Policy Studies[d] estimates that the World Bank spends 20 times more on projects involving fossil energy than it does on renewables. The think tank New Economics Foundation (nef)[e] found that no more than 3 percent of the funds for energy projects in developing countries involve renewables (out of a total of some $40 billion annually). At the same time, OECD countries spend up to six times as much on subsidies for fossil energy as they do on development aid.

According to the World Energy Council, renewables would become competitive very quickly if they received the subsidies that fossil energy and nuclear energy receive in one year: some 15 billion euros in the EU.[f]

a) See www.iea.org/Textbase/publications/free_new_Desc.asp?PUBS_ID=1263, cited Feb 12, 2006
b) "Two thirds of the energy subsides in EU-15 (almost 22 billion Euros) go to support fossil fuel production and consumption, while 1/6 (or 5 billion Euros) goes to support renewable energies." See http://org.eea.eu.int/documents/speeches/03-06-2004, cited Feb 12, 2006
c) See "A wrong turn from Rio," www.seen.org/PDFs/Wrong_turn_Rio.pdf, cited Feb 12, 2006
d) See "The World Bank and Fossil Fuels: At the Crossroads," www.seen.org/pages/reports/WB_brief_0903.shtml, cited Feb 12, 2006
e) See www.edie.net/news/news_story.asp?id=8519, cited Feb 12, 2006
f) See "The Subsidy Scandal: The European clash between environmental rhetoric and public spending," http://archive.greenpeace.org/comms/97/climate/eusub.html, cited Feb 12, 2006

While supporters of nuclear power seem to want to make people believe that Germany is the only country to phase out nuclear power, in fact Sweden, Austria, Belgium, the Netherlands, and Italy have also resolved to decommission plants rather than extend their service lives. Other countries, such as Great Britain, have not yet made a decision, but the British will have only one nuclear power plant online by 2025 if all their nuclear plants are decommissioned on schedule. And according to the 1999 *Dilemma Study*[2] by the EU Commission, in the business-as-usual scenario the share of nuclear power in the EU's electricity supply will drop from 23 percent in 1995 to 9 percent by 2025. By 2035, if nothing changes, the share of nuclear power in total electricity production in Europe will drop to 1 percent.

If we want to keep nuclear plants after that, uranium resources might be a problem. The IEA estimates current reserves at 50 to 60 years.[3] Or, as Helen Caldicott, president of the Nuclear Policy Research Institute, has stated, "If we decided today to replace all fossil-fuel-generated electricity with nuclear power, there would only be enough economically viable uranium to fuel the reactors for three to four years."[4] Others argue that reprocessing is the answer, but then we are dealing with plutonium, which can be used to make weapons, and the security risks only increase. The "breeders" that theoretically could produce more fuel than they consume have not proved to be a success. France closed its Superphénix because of cost overruns, and Japan closed its facility because of leaks.

The story of nuclear power in Great Britain is especially interesting. The privatization and deregulation of the electricity market began earlier there than elsewhere, and it became clearest there that nuclear power simply would not pay for itself in a free-market economy. Although nuclear power has been receiving massive subsidies for decades, British Energy has not been able to generate power at market prices in the past few years and had to be rescued with tax money because it was losing around one million euros per year. More than anything else, the free market showed that Britain had excess generation capacity.

The problem with getting out of nuclear power is "stranded costs." Most of the costs for nuclear power plants occur at the outset for construction and for provisions set aside for decommissioning. Fuel costs are relatively low compared to gas turbines, for instance. In other words, the longer a nuclear power plant runs, the more it pays for itself and if a nuclear plant is decommissioned ahead of schedule, these costs can never be recovered. In June 2003, researchers at MIT published the study *The Future of Nuclear Power,*[5] in which they explain: "Today, nuclear power is not an economically competitive choice." The EU's *Dilemma Study* came to the same conclusion: "Nuclear power is not generally perceived to have fulfilled the great expectations of its early days. It is marginally economic at best at present in most countries in Europe."

That has not stopped Finland from announcing that it will build a fifth nuclear power plant. However, when France announced in 2003 that it also had plans for a new nuclear plant, Greenpeace produced figures showing that the 3.5 billion euros the plant would cost would produce more energy if the money were invested instead in wind turbines: 24 terawatt-hours per year instead of 10.[6] In addition, Greenpeace found that the wind industry would create five times more jobs in this scenario than would a single new nuclear plant. And France still does not know what to do with its nuclear waste. Aside from the US, Finland is the only country in the world that has already specified the location of the final repository for its nuclear waste. Globally, 27 nuclear power plants are currently under construction, most of them in Asia.

In California, where deregulation got off to an early start, stranded costs were a major problem for nuclear power plants. Generally, taxpayers simply bought the nuclear plants so that deregulation could proceed. In Sacramento, for instance, the local utility SMUD decided to decommission an 800-megawatt nuclear power plant (Rancho Seco) in the mid-1990s.[7] Mostly because of stranded costs, SMUD had to pay $660 million to decommission it. In 1974, the power plant had only cost $342 million to build. Nevertheless, investments in the decommissioning of the plant paid off quickly. SMUD replaced the nuclear plant with efficiency, wind power, and solar power. Part of the solution seems too easy to be true: trees were planted to create more shade so that residents would not waste so much energy running air conditioners.

Germany's opting out of nuclear power revealed that it, too, had surplus generation capacity. Petra Uhlmann, spokesperson for E.On, the operator of the first of the two nuclear power plants shut down to date, commented on the occasion: "We would have shut down the power plant even without the agreement with the government."[8]

It would seem that nuclear power does not have much of a future. And yet, 2002 was a good year for the nuclear industry, with record production figures. Over the past few years, the nuclear industry has been using concerns about carbon dioxide emissions to its advantage. Christian Wilson, spokesperson for the German Atomic Forum in Berlin, recently explained: "With every kilowatt-hour of nuclear power, we avoid 1 kg of carbon dioxide in the atmosphere, which would have otherwise been created by less environmentally friendly sources of energy."[9]

This message is getting through to decision-makers such as Loyola de Palacio, former vice president of the EU Commission. As she put it, "Europe cannot renounce nuclear energy, not only for strategic reasons but also because of our commitments at Kyoto."[10] The MIT report *The Future of Nuclear Power*

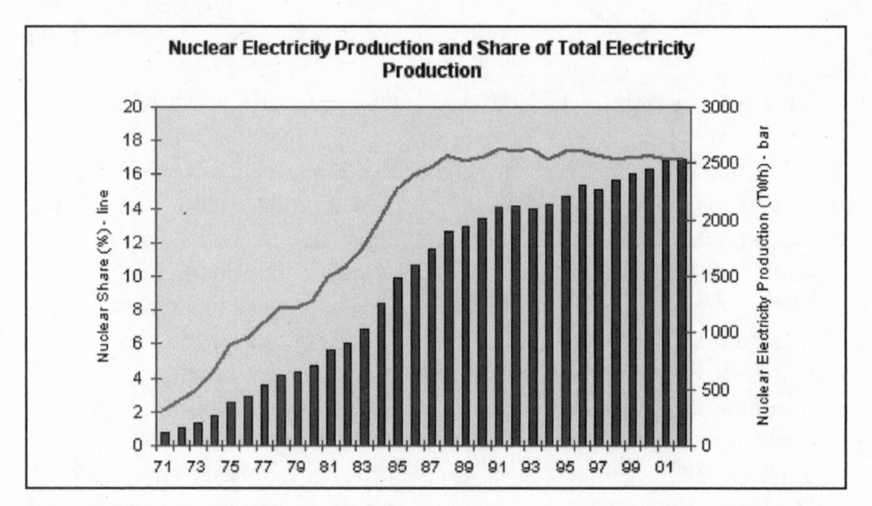

Figure 6.2:
The success story of nuclear power. The line shows the share of nuclear power in global electricity production. Since the late 1980s, this share has remained relatively stable at 17 percent. But the bars tell another story: production continues to increase, mostly because nuclear plants are running closer to capacity. In most places they cover the base load, which means they generally run full blast. In some countries, such as France, Belgium, and Lithuania, nuclear power makes up 60 to 80 percent of the electricity supply and nuclear plants have to be throttled when demand is low. (Source: World Nuclear Association)

sees things the same way: "We believe that the nuclear option should be retained, precisely because it is an important carbon-free source of power that can potentially make a significant contribution to future electricity supply."[11] Great Britain's energy minister, Stephen Timms, even went so far as to say that it doesn't matter if nuclear plants are not competitive; Britain will need them regardless if it is to meet its Kyoto targets.[12] But when US Vice President Dick Cheney complains because the government has not approved the construction of a single nuclear power plant for more than 20 years, we should not forget that no utility has filed for a new plant in the US during that time because nuclear power is too expensive. As Werner Brinker, president of the German Association of the Electricity Industry (VDEW), recently put it for Germany, "The investment costs for nuclear power plants are so great that no one is even thinking of building any."[13]

How Much Carbon Dioxide Will Nuclear Power Save Us?

In trying to determine emissions, a number of parameters have to be taken into consideration. For instance, are we talking only about carbon dioxide or about all heat-trapping gases? Are we talking about emissions during the generation of power or all the emissions from the construction of the power plant and the provision of the energy carrier (such as coal mining)? In addition,

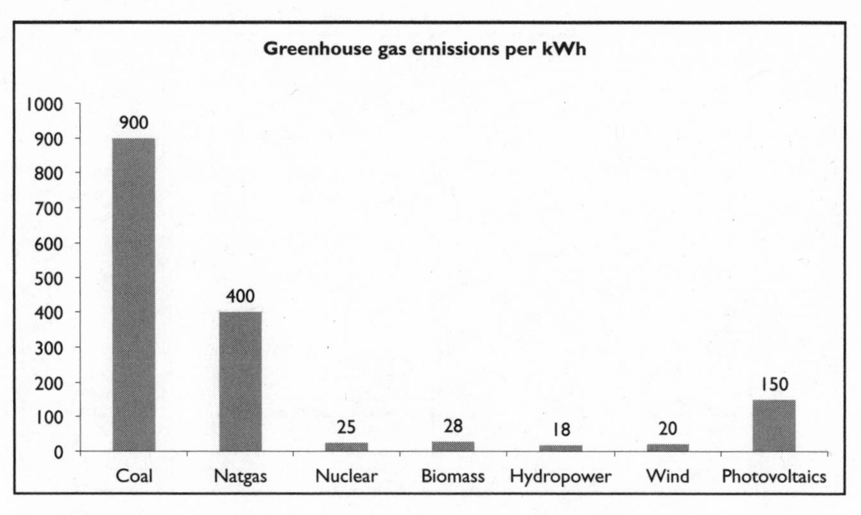

Figure 6.3:
Greenhouse gas emissions from various energy sources in grams per kilowatt-hour.
(Source: Terry Surles, California Energy Commission)

power plants are not all created equal. Emissions from coal plants vary greatly from one plant to another, with lignite (brown coal) generally being much "dirtier" than hard coal. Nonetheless, there can be no doubt that nuclear power is one of the winners when it comes to low emissions, as Figure 6.3 shows.

We occasionally come across higher figures for nuclear power, such as 39.1 grams of carbon dioxide per kilowatt-hour from a 1993 study for the Swiss Energy Ministry.[14] In that study the decommissioning of the power plant and the storage of nuclear waste were not included, but even if the figures were to double, they would still be excellent in comparison to coal, natural gas, and even photovoltaics. By way of comparison, Figure 6.4 takes into consideration plant construction, fuel provision, and plant operation.

With nuclear power, carbon dioxide is emitted mostly during the construction of the plant and the mining of uranium. As Figure 6.4 also shows, the various types of renewable energy entail emissions of carbon dioxide only for their manufacture. In other words, if these systems were manufactured using renewable energy, emissions of carbon dioxide for the construction of photovoltaic systems would be practically zero. Fossil fuels do not have to be used to produce solar panels, as Solar-Fabrik of Freiburg, Germany, demonstrates.[15] This company operates an emissions-free plant that gets its power from its own photovoltaics modules and a biomass cogeneration unit. The only emissions from this plant are related to purchased materials.

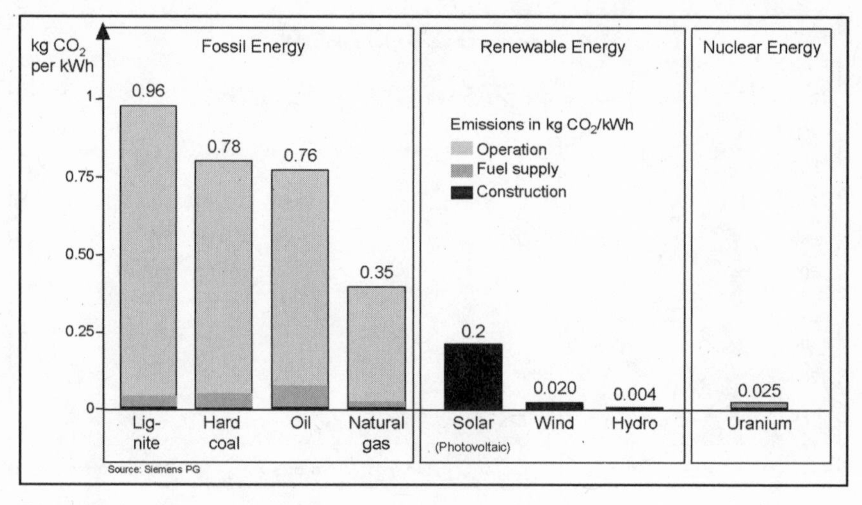

Figure 6.4:
CO_2 emissions by power source. (Source: Siemens)

Hearing Is Not Believing

Some environmentalists might not like to hear it, but nuclear power simply *is* an option for reducing emissions. The emissions from nuclear power plants are relatively low, even when compared to renewable energy. On the other hand, in light of the unsolved problems related to the long-term storage of nuclear waste and the danger of a meltdown, most people have a hard time understanding that nuclear energy can be considered "clean."

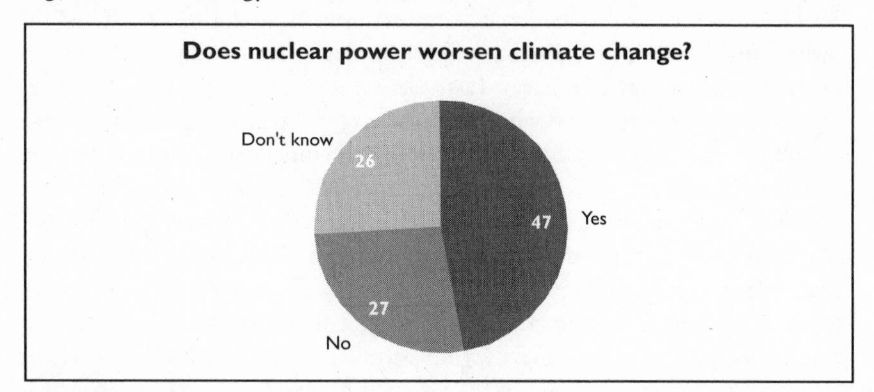

Figure 6.5:
Climate change and nuclear power. Almost half of the EU citizens surveyed in 2002 believed that nuclear power plants worsen global climate change, which is related to emissions of heat-trapping gases. Nuclear power plants do not emit these gases on a large scale. Ideologies aside, the people who answered "no" here are right. On the other hand, 93 percent of chlorofluorocarbons used in the US are for the enrichment of uranium. In addition to damaging the ozone layer, these gases also contribute to global warming. (Source: Eurobarometer 169 survey)

For instance, in the 2002 survey Eurobarometer 169,[16] EU citizens were not able to make a clear distinction between atomic waste (bad for the environment) and the low carbon dioxide emissions from atomic power plants (good for the environment). When asked which environmental problems worried them the most, 50 percent of those surveyed put nuclear disasters — not climate change — first. Nuclear power still has an uphill battle in selling itself as clean energy.

The situation is apparently not any different in the US, as we read in MIT's *The Future of Nuclear Power:*

> In the United States, people do not connect concern about global warming with carbon-free nuclear power. There is no difference in support for building more nuclear power plants between those who are very concerned about global warming and those who are not.[17]

This brings us to the ultimate argument of the nuclear industry, which was perhaps put best on the website of the World Nuclear Association in 2004: "'Renewables' like solar, wind and biomass can help. But only nuclear power offers clean, environmentally friendly energy on a massive scale."[18]

In other words, there is no alternative to nuclear power since fossil energy is dirty and renewables can only remain small. This line of reasoning is also at the heart of the criticism of Germany's phase-out of nuclear power. But is it true?

The Future of Nuclear Power: Not What You Are Thinking

Germany's success in replacing nuclear power with renewable energy is crucial because critics of renewable energy like to remind us of Sweden's failure some 25 years ago. In 1980, just after the nuclear accident at Three Mile Island, the Swedes resolved to shut down their 12 nuclear power plants by 2010. In 1986, in the wake of the disaster at Chernobyl, there was great public pressure to step up this plan. Sweden was to get its electricity from renewable energy and conservation. This goal turned out to be harder to reach than initially envisioned, and in 1995 the government ended up reconsidering shutting down the country's nuclear plants. Sweden will nonetheless have shut down all of its nuclear power plants by 2025 if they are decommissioned on schedule after 40 years of operation.

Does the example of Sweden not prove that renewables cannot replace nuclear power? Not at all, for the example of SMUD in California (discussed above) shows that energy efficiency and renewable energy were able to achieve this goal as early as 1995. On a similarly small scale, Austrians voted in 1978 not to build another nuclear power plant but to invest in renewable energy and efficiency instead. Back then, 86 percent of the citizens of the region of Vorarlberg voted against a nuclear power plant; today the region produces more green power than it consumes, exporting the rest.

Wind power has exceeded all expectations. A study conducted by Prognos AG in 1990 forecast in its status-quo scenario that 0.64 billion kilowatt-hours of wind power would be generated in Germany by 2005. In 2004, however, Germany generated over 26 billion kilowatt-hours of wind power — more than 40 times the forecast.[19] At present, the photovoltaics industry is booming in Germany, though it is starting from a very low level, just as the wind industry did in 1990. In addition, the first geothermal plants are just now being put into operation in Germany. In 2003, a geothermal facility in northern Germany started feeding power to the grid (see Chapter 10), and at the moment there is a lot of activity in geothermal energy in southern Germany.

The German government has undertaken to increase the share of renewable energy from just below 8 percent in 2005 to 12.5 percent by 2010. EU directive 2001/77/EEC specifies targets for 2010 for the EU-15 countries.

Targets for 2010 also have been set for eight of the EU accession countries.

Country	1997	2010
Austria	70%	78.1%
Belgium	1.1 %	6.0 %
Denmark	8.7 %	29%
Finland	44.7 %	31.5 %
France	15 %	21.0 %
Germany	4.5 %	12.5 %
Greece	8.6 %	20.1 %
Ireland	3.6 %	13.2 %
Italy	16%	25.0 %
Luxembourg	2.1%	5.7 %
Netherlands	3.5%	9.0%
Portugal	38.5%	39.0%
Spain	19.9%	29.4%
Sweden	49.1%	60%
UK	1.7%	10%
EU-15	13.9%	22%

Figure 6.6:
Targets for 2010 for the share of renewable energy in electricity generation in EU-15 countries. (Source: EU directive 2001/77/EEC)

Today, hydropower makes up a large share of the renewable energy in some countries (such as Sweden and Austria), but most of the growth from now until 2010 is going to come from wind power, photovoltaics, and biomass. In Great Britain, offshore wind is expected to boom and there is great potential for tidal power plants and wave power. By 2020, five years before Germany shuts down its last nuclear power plants, the German government aims to have a 20 percent

Country	Share of renewable energy in the generation of electricity by 2010
Latvia	49.3 %
Slovakia	33.6 %
Slovenia	31.0 %
Czech Republic	8%
Poland	7.5 %
Lithuania	7%
Estonia	5.1 %
Hungary	3.6 %

Figure 6.7:
Share of renewables in the 10 Accession Countries to EU-25.

share of renewable energy — almost 8 percent more than in 2010. Currently, nuclear power provides less than 30 percent of Germany's electricity, but by 2020 renewable energy will be covering an additional 12 percent of the pie. In other words, in 2020 we will have five years to fill an 18 percent gap at growth rates for renewable energy of 1 percent annually.

Can Renewables Ever Provide 100 Percent of Our Supply?

This status-quo scenario still leaves us with a fairly large amount of power to account for. However, it is certainly feasible to close that gap by increasing efficiency and conserving, which we will come back to in later chapters. Suffice it to say here that it is not only environmental groups who see incredible savings potential. In 2003, Esso (the European subsidiary of ExxonMobil) forecast a 7 percent drop in the consumption of primary energy in Germany by 2020, partly due to the aging population.[20] But simply stepping up the use of cogeneration plants, for instance, would allow Germany to get much of its heating from the waste heat created when power is generated. Proponents of cogeneration units such as Johannes van Bergen, director of municipal works of the town of Schwäbisch-Hall, thus cannot understand why people are so skeptical that we can do without nuclear power:

> The state governments in Germany have already taken it upon themselves to double the share of renewable energy by 2012. In addition, expert reports contracted by the state governments have found that the share of cogeneration in the supply of electricity can be as great as 30%. It thus does not make any sense that the current agreements between the industry and the federal government to shut down nuclear power plants are being called into question.[21]

Indeed, we can take this line of thinking even further. Globally, a complete switch to renewables is feasible — in other words, in the long-term we can do not only without nuclear power but also without coal plants and gas turbines, replacing them all with renewable energy.

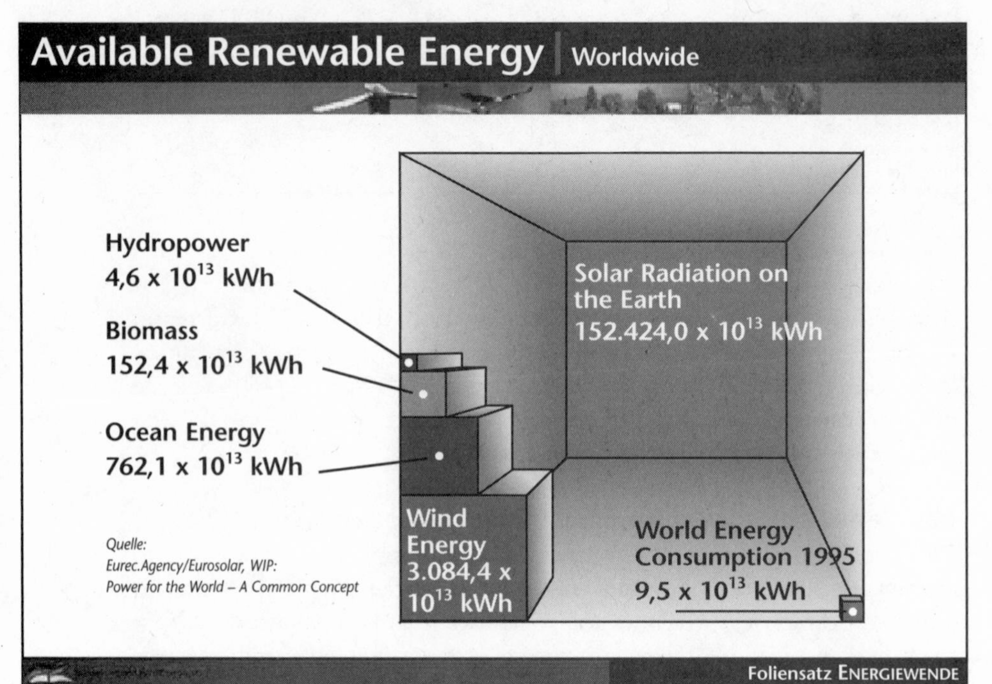

Figure 6.8:
Potential of renewables. The potential of hydropower is equivalent to 50 percent of total global energy consumption in 1995. In contrast, biomass could theoretically provide 16 times more energy than was consumed that year, while wave and tidal energy could produce 80 times as much, wind 325 times as much, and solar energy an unbelievable 16,044 times as much. These figures are purely theoretical but they show that renewables do not have to remain a niche product. There are further interesting figures. For instance, the British firm Cyberium calculated that the annual solar radiation incident on London and its suburbs is equivalent to the annual global consumption of fossil energy. Others have calculated that the energy the sun sends to Earth in one hour is twice as great as annual global energy consumption. (Source: Eurec Agency)

However, the potentials mentioned in Figure 6.8 are completely theoretical. For instance, the inefficiency of power generators would reduce these figures drastically. The greatest theoretical efficiency that a wind turbine can reach has been calculated to be 59 percent (the Betz limit), but in practice the values are far below even that. Over the course of a year, wind turbines in Germany run on the average at 17 percent of their rated capacity. That would reduce the potential for wind energy to around one-sixth of 325 times the energy consumed in 1995, or roughly 54 times that amount. Of course, then the landscape and the seascape would be covered with wind turbines, which nobody wants. But then again, we don't need 54 times our current energy consumption from wind. Maybe 20 percent will do.

Even if we take the much more conservative figures from Worldwatch Institute, we still have a lot of leeway. Christopher Flavin of Worldwatch has calculated that under realistic conditions the world's wind energy potential is five times greater than its energy consumption, and that the US could get 20 percent of its power from wind turbines in just 0.6 percent of the country.[22] And according to the statistics of the US Public Interest Research Group Education Fund, the wind energy in only four states — North Dakota, South Dakota, Kansas, and Nebraska — would suffice to cover 100 percent of the power consumption in the US.[23] The US Department of Energy has also calculated that just 9 percent of the solar energy in Nevada would cover power consumption in the whole country.[24]

But again, let us not be confused by such statistics. The goal is not to supply energy to a whole country from some small area. This is why studies that show what the different types of renewable energy can contribute to a country's energy supply are so much more important than the theoretical potential of one source. For instance, the report *Energy Rich Japan,* published in 2003 by the Wuppertal Institute of Germany, shows that even a country as poor in resources as Japan can completely switch over to renewable energy. And, as the authors are quick to point out, if Japan can do it, then just about any other country can:

> If it is possible to achieve a 100% renewable energy system in Japan using today's best available technology, it would be possible to transfer and adapt the results to many other locations even to cover the whole globe.[25]

Twenty percent wind energy (as Flavin calculates) could indeed be enough if the other 80 percent came from sources such as photovoltaics, geothermal, biomass, tidal power, and wave power. Pollution would then be largely a thing of the past and there would be no risk of a nuclear meltdown. Even the proponents of nuclear power admit that nuclear power plants are not completely risk-free. They speak of a "residual risk." In the December 2003 issue of the German edition of MIT's *Technology Review,* Adolf Birkhofer, who assesses the risks of nuclear power plants in Germany, estimated the risk of accidents happening if the world has 500 nuclear power plants in the future (compared to the approximately 440 today): "Every 200 years, there would be a major accident somewhere in the world."[26] So if a mixture of different types of renewable energy can provide all the energy we need, would anyone want to run that residual risk of nuclear power? At that point, does nuclear power offer any benefits?

Power on Demand

Nuclear power plants offer only one benefit: power on demand (after a certain run-up phase), even in large mobile applications. Wind turbines and solar panels generate power only when the wind is blowing and the sun is shining. But that drawback is less problematic than most people think. After all, as we know from a number of off-grid projects and microgrids, the sun generally shines when the wind is not blowing and vice versa. In addition, we can focus more on demand management so that we consume more power when it is being generated (see Chapter 13).

The only types of renewable energy that provide power on demand are biomass, geothermal, and hydropower. Of these three, only biomass is able to produce power in mobile applications, such as in ships. Here, nuclear power has an advantage. Indeed, a lot of attention is currently being paid to the development of "small" nuclear power generators. While a nuclear power plant commonly has a capacity of around 1,000 megawatts, models with 200 megawatts and less are being further developed. Why?

There is a consensus that large, central power plants are wasteful. Huge amounts of power are produced in one area and then have to be transported to consumers elsewhere. In France, almost 6 percent of the power generated is lost in transport because the country gets almost 80 percent of its electricity from large, central nuclear plants. In contrast, Germany loses only around 4 percent of its power in transport with its slightly more distributed power generation infrastructure. According to one estimate by the multi-agency US Climate Change Technology Program,[27] distribution and transport losses in the US amount to only 7.2 percent, which is higher than in France or Germany but is nonetheless impressive given the much longer distances covered in the US and the age of the grid, which was greatly expanded in the 1930s under Roosevelt's rural electrification plan. In addition to reducing power loss, when the units are close to consumers the waste heat from small generators can be put to good use, such as in district heating networks.

The Japanese are even working on a 10-megawatt nuclear power plant that will be a tower roughly the size of a fully grown pine tree. Such nuclear plants can power a wide range of applications, including ships. The first "floating" nuclear plant was installed on the USS *Sturgis,* which was stationed in the Panama Canal. Its job was to provide 45 megawatts of power for the operation of the canal. Such ships can be moved around quickly and would provide enough power for a small coastal town. Russia now aims to take advantage of this possibility to get around international law. Russia has 70-megawatt nuclear generators that it uses on its icebreakers. It wants to install such plants on ships that will be anchored off the coast of India starting in 2008. The power generated from the ships will be fed to the Indian grid. This approach will allow power to be sold to India, but the nuclear plants will remain in Russian hands. Russia will thus be able to sell nuclear power to India without having to sell a nuclear plant. Russia is forbidden by international law from selling nuclear power plants to India because India is not a member of the Nuclear Suppliers Group. (The Bush administration simply ignored this law in 2005 when it announced that the US would share nuclear technology with India in flagrant violation of the Non-Proliferation Treaty.)

The Secret Life of Nuclear Power

Nuclear power — not just nuclear weapons — has great military potential. When oil has become scarce, with what will we fuel our aircraft carriers? Already, the French aircraft carrier *Charles de Gaulle* is powered with two nuclear generators, each with a capacity of 61 megawatts. Countries that want to be able to deal on an equal footing with the United States militarily in 50 years will need not only nuclear weapons but probably also nuclear power. The US military is currently very interested in the manufacture of hydrogen from nuclear power to refuel military vehicles running on hydrogen-powered fuel cells.

It is going to be difficult to wage war with solar and wind power but that does not mean that an age of everlasting peace will dawn once we run out of oil. Rather, nuclear energy will be an important source of power for modern military machines. In light of the recent US tendency to launch preemptive strikes, opponents of nuclear power outside the US who also are interested in world peace based on civil negotiations between equal nations will have to ask themselves whether they are willing to leave nuclear weapons to the countries that currently have them and whether nuclear power might have a role to play in deterring aggression from the world's only remaining superpower. In addition, Americans interested in peace should ask themselves whether the US can afford to continue to force countries to go nuclear before the US treats them with respect.

Out of the Frying Pan: Green Guru James Lovelock Calls for an Expansion of Nuclear Power

James Lovelock popularized the concept of Gaia: the Earth as a living organism that regulates itself. In 2004, many of his followers were shocked to hear then 84-year-old Lovelock call for more nuclear power to battle global warming.

Of course, nuclear power is very "clean" in terms of pollution and emissions of greenhouse gases, as long as we forget about nuclear waste or consider it relatively harmless or at least manageable. As we have seen, those who support nuclear power not only describe it as clean but also claim that renewable energy will remain comparatively irrelevant. And this is exactly the line of thinking that James Lovelock has fallen prey to:

> We cannot continue drawing energy from fossil fuels and there is no chance that the renewables, wind, tide and water power can provide enough energy in time. If we had 50 years or more we might make these our main sources. But we do not have 50 years; the Earth is already so disabled by the insidious poison of greenhouse gases that even if we stop all fossil fuel burning immediately, the consequences of what we have already done will last for 1,000 years. Every year that we continue burning carbon makes it worse for our descendants and for civilisation.[28]

Of course, what many people have overlooked in Lovelock's argument is his opposition to coal power (see Chapter 5). After all, Lovelock knows exactly how the growing concentration of carbon dioxide in the atmosphere will drastically change the global climate, with devastating effects on humankind.

While it is possible that some areas (such as Russia and Germany) theoretically could benefit from slightly higher temperatures, we do not have any idea what we are getting ourselves into. Would slightly higher temperatures allow plants to grow faster and would the increase in vegetation bind more carbon dioxide, thus keeping the increase in greenhouse gases in check? Then we might even have greater biodiversity and better harvests, as some researchers are predicting. Or would the relatively cold Germany have a climate more like that of Oklahoma, as one of Germany's most prominent climate researchers believes? Or would Europe fall back into an Ice Age like the one described in the movie *The Day After Tomorrow*?

All we can be sure of is that we cannot prepare for fast changes if we don't know what they will be. If I know that Germany is going to have better harvests, I will plant fruit trees. But if Germany is going to be a desert in 20 years, I'm going to sell my house and move away (but to where?). Precisely because we cannot predict anything — and, of course, because people cannot simply get up and move where they want internationally — any drastic climate change has the potential to destabilize civilization.

Lovelock understands climate mechanisms probably better than anyone else. But what are we to make of his proposal to use nuclear power to save the Earth? Here, the congregation is quickly divided into the believers and the non-believers: declare nuclear waste manageable or avoid it at any price. Lovelock is of the former persuasion:

> Opposition to nuclear energy is based on irrational fear fed by Hollywood-style fiction, the Green lobbies and the media. These fears are unjustified, and nuclear energy from its start in 1952 has proved to be the safest of all energy sources. We must stop fretting over the minute statistical risks of cancer from chemicals or radiation. Nearly one third of us will die of cancer anyway, mainly because we breathe air laden with that all-pervasive carcinogen, oxygen.[29]

It would seem that we will have to wait until the next nuclear catastrophe before people will stop arguing along these lines for at least a few years. Many people probably thought that Chernobyl was enough, but even today there is no consensus about how many people have died from the effects of nuclear power. The estimates for Chernobyl alone are as high as 15,000. In 2003, the International Atomic Energy Agency created a Chernobyl Forum to finally clear up this issue. The Forum's full report is expected in 2006.

In comparison, wind expert Paul Gipe keeps a count on his website[30] of the number of people that wind energy has killed. Except for one paraglider, all 22 people died during construction or maintenance work. In other words, no one has been killed, say, from a falling turbine. (We will leave it up to the reader to decide whether the death of the paraglider should be attributed to wind power or to paragliding.) That is not very many people if we keep in mind that the construction of the 2,000-megawatt Hoover Dam led to the deaths of some 100 people.

When it comes to renewables, proponents of nuclear power can claim only that renewables are simply too small to be counted. But it ain't so. Today, no one can claim that wind power is small. The installed capacity worldwide far exceeds 40,000 megawatts — 20 times greater than that of the Hoover Dam. In addition, it is not at all clear that nuclear power can be expanded faster than renewables. Nowadays, all kinds of renewable sources of energy are booming worldwide (photovoltaics at a rate of around 30 percent annually), while it would take eight to 15 years to build a nuclear plant. The International Atomic Energy Agency contradicted Lovelock's claim that nuclear power could save us from climate change: "Climate change would doom the planet before nuclear power could save it."[31]

So when Lovelock writes that we don't have another 50 years and we have to act quickly, we have to agree. But the fact remains that wind turbines, photovoltaics, and other renewables can be developed faster than nuclear power. They just need to receive the support that fossil energy and nuclear power (at least up until Chernobyl) have received. There is no denying that the funding for fossil energy and nuclear power has greatly exceeded the funding for renewables. According to the US Public Interest Research Group, in the US some three times as much has been spent in the last 50 years on fossil energy and nuclear power as on renewables.[32]

Of course, we could expand both renewables and nuclear power at the same time — one does not rule the other out. One might even argue that they complement each other wonderfully: nuclear power covering the base load, with biomass, geothermal, and hydropower compensating for fluctuations in wind and solar power. Of course, if we increase efficiency considerably, we won't have to do much expansion at all.

All of this could help mitigate the effects of global warming in the long term. The environment will be grateful, especially if turbines are not put up in sensitive preservation areas or too close to buildings. In Germany, such turbines generally have to be 300 to 1,000 meters from the next building to make sure that no one is bothered by the noise from a wind turbine or by the shadows cast by the rotating blades. At the same time, nuclear power plants also should not be built too close to other buildings just in case there is an accident (which, admittedly, rarely happens). In Chernobyl, the quarantined area was around 2,800 square kilometers, though the whole area that received large amounts of radiation was about the size of Switzerland. In addition, nuclear power plants should not be built in areas prone to natural disasters like earthquakes, such as Japan, which has 52 plants. As Leuren Moret, the whistle-blower from the Lawrence Livermore Nuclear Weapons Laboratory's Yucca Mountain Project, recently put it, "Of all the places in all the world where no one in their right mind would build scores of nuclear power plants, Japan would be pretty near the top of the list."[33]

So where does that leave us? No nuclear power plants in Japan, on the Rhine, on the Mississippi. If I had the choice, I would prefer to pass on to my children a clean planet that may be studded with wind turbines, which can be dismantled at any time without leaving a trace, but not with such time bombs as nuclear power plants and nuclear waste storage facilities.

Yucky Yucca

In February 2002, George W. Bush approved the final repository for nuclear waste from all over the country in Yucca Mountain, Nevada. Soon afterward, the governor of Nevada vetoed the project — after all, who wants nuclear waste? The US Senate overruled his veto and gave its approval for the final repository of highly radioactive waste from 131 nuclear power plants in 39 states. Some 100 miles from Las Vegas, Yucca Mountain is to become the final resting place for nuclear waste made in the USA.

What else are we to do with our nuclear waste? Back in 1957, the US government had to start dealing with that question when the first nuclear plant was put into operation. Should we just shoot the waste into space? It is only a matter of time until one of the rockets blows up. Why not just drop it into the ocean? Even in the 21st century, we know little about the floor of the ocean, and US law stipulates that the waste has to be safely stored away for at least 10,000 years. Sounds like we will have to bury it somewhere.

The US Department of Energy chose Yucca Mountain as the best of three proposed sites. Even before the site was approved, billions of dollars had been invested in the project. A tunnel had been dug some five miles into the mountain for test purposes. Perhaps appropriately, Yucca Mountain is near the Funeral Mountains and Death Valley. It is also on government land close to where atomic bombs were tested for some 45 years, most recently in 1992.

Of course, the mandatory 10,000 years of safe storage is not sufficient, since some radioactive waste will remain dangerous for hundreds of thousands of years. But this kind of understatement has a long tradition in the nuclear industry. Remember the old "duck and cover" public service announcements? They were intended to give people the impression that they could do something to protect themselves in case of an atomic attack. The Price-Anderson Act of 1957 also limited the liability of operators of nuclear power plants to $200 million per plant (with another $88 million added later), although the US government found in 1982 that the damage could easily reach $560 *billion*. Incidentally, the Price-Anderson Act was extended on November 28, 2001 for another 15 years.

If you think that the costs for the storage of nuclear waste are included in the price of the power generated, think again. Of the $460 million that the government invested in the Yucca Mountain project in 2002, only $145 million came from the budgets of power plant operators. One-tenth of 1 cent per kilowatt-hour of nuclear power is invested in the Nuclear Waste Fund. At the beginning of 2004, the fund had some $24 billion.[34] In comparison, the costs for the repository at Yucca Mountain are estimated at around $60 billion.

The Department of Energy says that the seven volcanoes near Yucca Mountain will "probably not" erupt in the next 10,000 years. Interestingly, the DOE estimates that the radiation emitted from Yucca Mountain will increase over time because "somewhere between 1,000 and 10,000 years from now, the climate is anticipated to begin changing."[35]

Shifting Priorities

Beyond a doubt, Yucca Mountain is attractive not only because it is an old test site for atomic bombs and is already owned by the DOE but also because it almost never rains there. Groundwater is one of the greatest dangers to the nuclear waste containers. But Yucca Mountain also has a major drawback: it is in an area where earthquakes are not uncommon. The earthquake that registered 4.4 on the Richter scale on the morning of June 14, 2002 — just a few weeks before the Senate approved the repository — woke up many of the residents in the region but did not stop Congress from approving the site. The DOE was quick to issue a press release on June 14 stating: "There was not an earthquake at Yucca Mountain today. There was a 'light' earthquake located at Little Skull Mountain, Nevada, as reported by the U.S. Geological Survey early this morning. According to the USGS, the quake's epicenter was located approximately 15 miles to the east of Yucca Mountain." [36]

In the same press release, the DOE reassured everyone that not even the earthquake that registered 5.6 on the Richter scale in 1992 would have been able to damage the repository. The official website for the project also claims that the repository could withstand even more severe earthquakes. The choice of Yucca Mountain was politically motivated. As Alison McFarland of the Massachusetts Institute of Technology told *The New York Times:* "The politically weakest state under consideration ended up with the repository."[37]

But potential dangers do not matter much when the intermediate repositories are full. After all, we can't switch off nuclear plants overnight just because we don't know what to do with their waste.

Mobile Chernobyl

The residents of Nevada are not the only ones who opposed the project. At the moment, more than 50,000 tons of nuclear waste are being stored in interim sites in 39 states. An estimated 161 million people live within 75 miles of at least one of these sites. Websites have been created where people can enter their location and find out where the nearest route is for the transport of nuclear waste from an intermediate site to the final repository.

Starting in 2010, more than 10,000 trains and more than 50,000 trucks could be transporting nuclear waste across the US over 24 years. It is still not clear how all this will work because, while Yucca Mountain itself has been studied intensely, less attention has been paid to transporting the waste. The DOE states that there will be 175 deliveries per year, amounting to a total of some 4,200 deliveries by "dedicated trains." [38] But some critics have estimated that there may be as many as 76,000 trucks and 10,000 trains transporting nuclear waste in mixed transport. [39] In turn, the DOE points out that nuclear waste has already been transported some 2,700 times in the US over the past 30 years without any accidents. [40]

Regardless of whether transport is done by train or truck, 77,000 tons (more than 3,000 tons per year) of highly radioactive nuclear waste will be rolling through the country. In comparison, when the World Trade Center fell, some 60,000 tons of rubble had to be cleared up. And yet, Yucca Mountain will not even be able to store all the nuclear waste that will be created in the US by 2034. We will need another repository just to store the waste from the plants we already have.

In addition to terrorist attacks on trains and trucks transporting the waste, possible long-term damage from earthquakes and volcanoes close to the mountain, and accidents during transportation, many people are afraid of the radiation from passing trains and trucks. But the Nuclear Energy Institute reassures us: "You would receive as much radiation from eating bananas as would a pedestrian watching a year's worth of used nuclear fuel shipments pass by."[41]

That certainly is reassuring. But how is a country that may not even exist in another 200 years going to guarantee that no one in the next 10,000 years will suffer from this radioactive waste? Should we put up warning signs? And if so, which symbols will be understood regardless of language and culture? How can we make sure that the symbols we use will warn people instead of attracting them? The debate goes on.

Proponents of the repository at Yucca Mountain are right about one thing: we can't leave the waste sitting around forever at interim storage sites, which often are dangerously close to human settlements. The debate about Yucca Mountain has made it clear that if we don't get a final repository soon, all of the nuclear power plants in the US will have to be shut down.

That does not mean that we have a solution for the long-term storage of nuclear waste. Rather, we are simply pretending that Yucca Mountain is one. The people who claim today that this solution is safe will not be here long enough to guarantee future safety. Indeed, they will probably not be around long enough to suffer if something goes wrong. Our children and grandchildren will not thank us for this heritage.

Chapter 7

Natural Gas

More than 100 billion cubic meters of natural gas are flared around the world each year .... Were the flaring of gas in Africa alone to be used for power generation in efficient power plants, this could produce approximately 50 percent of the current power consumption of the African continent.

—Peter Woicke, managing director of the World Bank, at the World Summit on Sustainable Development, Johannesburg, 2002

In 1859, oil was discovered in Pennsylvania. Back then, people were looking for kerosene to replace whale fat, which had become expensive. Unfortunately, only about 10 percent of petroleum can be refined into kerosene. The other 90 percent was largely a waste product for which there was little or no demand. About half of petroleum can be refined to make gasoline and another fourth can be turned into diesel/heating oil. The latter could have been used immediately to replace the vegetable oils being used to lubricate machines. A combustion engine that could be powered with gasoline was invented in 1876. The first diesel engine, built by German Rudolf Diesel in 1892, was designed to run on peanut oil. Only later was the fossil fuel that could be used in his engines named after him as well.

For centuries, natural gas was merely flared off as a waste product at oil fields to reduce the danger of explosion. At the beginning of the 20th century, during the heyday of coal and before electricity had truly taken off, many cities had street lighting that burned a synthetic gas made from coal ("town gas"). Only after World War II, when oil had started to become scarce, was natural gas gradually considered as valuable as gasoline, diesel, and kerosene.

Since then, a lot has changed. Natural gas consumption grew globally by around 75 percent from 1985 to 2005 but a lot of natural gas is still wasted. The 100 billion cubic meters of flared natural gas that Peter Woicke spoke of at the Johannesburg summit is roughly the amount of natural gas that Germany con-

Figure 7.1:
In addition to European conglomerations in the Netherlands, the Ruhr Area in Germany, the belt from London to Manchester, and the Po Valley in Italy, this satellite photo of the Earth by night displays some very bright areas in the middle of the North Sea, where no one lives but plenty of natural gas is flared off. If we then look at the Persian Gulf, the oil fields are far brighter than the cities, and in Nigeria the cities of Lagos and Ibadan, each with millions of inhabitants, appear as tiny dots next to the giant oil fields east of them. (Photo courtesy of NASA)

sumes annually, equivalent to 2 percent of annual global petroleum consumption. Flared natural gas also makes up 1 percent of global carbon dioxide emissions.

For these reasons alone, if we are going to burn it anyway it makes good environmental sense to use natural gas to provide energy as a fuel for cars, for heating, for cooking, or to generate electricity. Natural gas is relatively environmentally friendly compared to coal and oil. No wonder such "green" thinkers such as Germany's Wuppertal Institute are calling on people to start buying cars that run on natural gas. Indeed, Germany's former environmental minister, Jürgen Trittin, Green party member, has repeatedly spoken of "environmentally friendly natural gas."[1]

The Greens Support a Form of Fossil Energy?

Trittin was referring to steam turbines fired with natural gas. They have efficiencies of up to 70 percent because they operate as cogeneration units, meaning that the waste heat is used as a source of energy instead of simply being released back into the atmosphere or neighboring bodies of water, as is the case with nuclear and coal plants. In addition, natural gas turbines can be flexibly scaled: from giant power plants with an output exceeding 1,000 megawatts (similar in size to a large nuclear or coal plant) down to the new microturbines with an output of around 25 kilowatts. Microturbines are a good solution for the problem of flaring because they can be used directly on oil platforms to provide power where the grid cannot reach.

Gas turbines not only can be dimensioned to suit specific needs but also can be run up and down to meet fluctuations in power demand much more quickly and flexibly than central nuclear or coal plants. In addition, the output of such turbines can be adjusted without making the units uneconomical because most of the costs of operation are related to fuel consumption. Large coal and nuclear plants have to run at high capacity to be economical.

The technology behind these cogeneration gas turbines came from the jet engines used in airplanes, which is why these turbines are also referred to as aeroturbines. In an energy supply system with a large share of intermittent renewables, such turbines will be useful to compensate for fluctuations in the supply of renewable energy. They are popular among environmentalists because the carbon dioxide emissions are only around 40 percent as high as from coal plants and only 50 percent as high as from oil-fired plants. Such conventional fossil plants generally have efficiencies of around 30 to 40 percent, far below the 60 to 70 percent efficiency of natural gas turbines.

Nonetheless: Finite Resources

The main reason that natural gas has been so successful in the US since 1990 is economics. In the 1990s, around 90 percent of newly installed electricity generating capacity was natural gas turbines. That was to have continued to around 2010, but in the past few years the supply of natural gas has been tight and prices have risen. Canada has great reserves, but the US does not, which is why the US is looking for ways to expand its liquefied natural gas (LNG) infrastructure to allow for increased imports of natural gas from overseas. Part of Bush's energy plan in 2005 was devoted to expanding LNG ports — ironically as part of an effort to reduce the country's energy dependence.

In Europe, pipelines can be used to bring natural gas west from Russia without having to liquefy it for transport. Russia has the largest natural gas reserves in the world — roughly a third of global reserves — and has signed supply contracts with Germany (among others) for the next 20 years. (If you are wondering why Germany is so reluctant to criticize Russia's actions in Chechnya, look no further.)

Worldwide, natural gas has been booming. Countries such as Spain and South Korea have been posting two-figure growth rates in the industry for many years. Currently, South Korea imports more LNG than any other country except the US. The study *World Energy, Technology and climate policy Outlook — 2030,*[2] conducted for the European Commission, states that natural gas will make up 20 percent of the EU's power supply by 2010. In 1990, it made up only 12 percent.

Current estimates are that the world's natural gas reserves will last for another 60 years at current rates of consumption — a full 20 years longer than current oil reserves. Unfortunately, natural gas reserves are not equally spread across the planet. As demand increases, countries like the US with high demand and low supply are likely to insist that other countries hand over their resources on the "free" market. In 2004, we saw what the effects of this situation can be when violence erupted in Bolivia, where citizens were protesting the sale of the country's resources to the US company Sempra. Sempra is a major player on the natural gas market, having built two of the largest, most modern cogeneration plants in the world that are fired with natural gas. The poor people in Bolivia were not about to stand by and watch a rich country take their resources away. In contrast, renewable energy is cleaner than natural gas and is available in sufficient quantities everywhere.

Not only are natural gas reserves unequally distributed but the fields also often are small and far from consumers and energy distribution infrastructure. Natural gas has a comparatively low energy density. A mass of natural gas contains only about a quarter of the energy of an equivalent mass of oil. As a result, people have been looking for ways to make it more profitable to transport natural gas. Until a few decades ago, many natural gas fields were considered "stranded"; the costs for transport would have made it uneconomical to market the gas. After all, a pipeline can easily cost a million dollars per kilometer.

One common way of solving this problem has been to convert natural gas into a synthetic fluid fuel on location. Ninety percent of natural gas is methane, which can be turned into diesel to make it easy to transport without great losses. Unfortunately, this process is not quite competitive economically. Recently, the industry has been cooling the natural gas down to -164 degrees Celsius, at which point the gas turns into a fluid: liquefied natural gas. The energy density of LNG is three times that of natural gas, which means that LNG contains around 75 percent as much energy as the equivalent mass of gasoline.

If we want to fill up our cars with LNG or natural gas, we are going to have to fill up the tank more often. One way of getting around this problem is to use natural gas vehicles in fleets that do not travel great distances from a hub (such as buses and taxis), a practice that is becoming more popular in Europe.

Natural gas cars have become popular in Argentina, which not only recently went through an economic crisis that made energy imports prohibitively expensive but also has the second-largest natural gas reserves in South America (behind Venezuela). Depending on taxation, natural gas can be less expensive than gasoline. In Europe, a full tank of natural gas costs half as much as a tank of gasoline.

LNG has one a major drawback, however: the energy needed to cool it down for liquefaction is equivalent to around 25 percent of the energy in the natural gas itself. In other words, transport losses start off at 25 percent before the LNG has even been moved. Nonetheless, according to a 2003 study by Germany's Wuppertal Institute,[3] these losses are less than the losses that occur when natural gas is converted into synthetic diesel. In many cases, such as in tankers, liquefaction truly pays for itself because the volume of liquefied natural gas is 1/600 that of natural gas. However, such vessels are accidents waiting to happen, as the world witnessed in April 2004 following the train accident in North Korea involving LNG.

Methane Hydrates

In the end, natural gas is still a form of fossil energy that contributes to global warming and pollutes the environment (though less than oil and coal). After all, natural gas is basically methane, and methane is a heat-trapping gas whose effect on global warming is 21 times greater than carbon dioxide's. Flaring the gas off converts it into water and carbon dioxide, which actually reduces the effect on global warming. In other words, what we want to prevent even more than flaring is methane being released directly into the atmosphere.

The largest methane reserves by far occur in a form that researchers have not yet been able to find a way to market: frozen balls of methane. Such "hydrates" are found on the ocean floor and in frozen tundra. They remain frozen up to around 18 degrees Celsius, at which point the methane is released into the atmosphere. These frozen blocks can be set on fire, which is why they are also referred to as "burning ice."

As a source of energy, these hydrates are available in greater quantities than any other source of fossil energy. It is estimated that they alone contain twice as much energy as all the oil, coal, and natural gas reserves the Earth has ever had. The US alone is estimated to have 33 times more hydrates than the current worldwide reserves of natural gas.[4] The problem is how to find a way to use these hydrates in the existing natural gas infrastructure. One cubic meter of methane hydrates contains about 164 cubic meters of natural gas. However, if we start burning this methane, we will release even more heat-trapping gases into the atmosphere.

We may be stuck between a rock and a hard place, for leaving the hydrates in the ground might not save us from these conditions. If the climate heats up as expected, at some point the first wave of these hydrates will begin to melt, releasing a tremendous amount of methane into the atmosphere and drastically accelerating the warming process, possibly until almost all of the world's methane hydrates have melted. The Earth would be a far warmer place indeed.

The process may already be underway. Researchers from Oxford University announced in 2005 that a section of Siberia larger than France was thawing.[5] The permafrost could annually release roughly the same amount of methane into the atmosphere as the planet's wetlands and agriculture do.

Such a scenario is not, unfortunately, a doomsday vision of environmentalists who are out of touch with reality. In fact, something very similar happened 55 million years ago and the result was a wave of mass extinction. In 2004, the head scientist of the British Climate Group warned that this tragedy could repeat itself this century.[6] The Earth would then be seven degrees hotter and would have no ice. In that case, the Antarctic is the place you would want to be.

Natural Gas as a Bridge

In addition to the argument that it makes sense to use natural gas as a source of energy instead of simply flaring it off, there is one more reason to use it: as a bridge in the transitional phase on the path to renewables.

On the one hand, if we are going to have the hyped-up hydrogen economy, people will have to get used to dealing with a gaseous form of energy instead of fluids and solids. Filling up your tank will be a little bit different, you will not be able to drive as far, and the dangers will be slightly different (flames at accidents may be above your head rather than on the ground). On the other hand, natural gas itself is now the main source of hydrogen for fuel cells, and the infrastructure we will need for hydrogen probably will be based on the one we now have for natural gas.

The methane in coal mines also is playing a role in the development of fuel cells. When water is pumped out of the mines, pressure below ground drops, releasing the methane in the coal and sometimes causing an explosion. Until recently, mining companies were more than happy to let much of this gas escape unused in order to make conditions for coal miners safer. But now more and more fuel cells are generating electricity from mine gas.

Finally, the hydrogen economy seems to be the ideal solution for one of the potential problems with renewable energy that has yet to be solved: how we will store excess energy in the distant future if we produce more electricity from renewables than we can consume at that moment. Natural gas could indirectly help us find out whether hydrogen can be a large-scale storage medium.

Chapter 8

Photovoltaics

Photovoltaics is the most promising of all types of renewable energy globally.
—Hermann Scheer, member of the German parliament and winner of the 1999 Alternative Nobel Prize for his work to promote renewables

Why Cloudy Germany is Making the Global PV Market Boom

When we hear the words "solar energy," most of us probably think of solar cells, also known as photovoltaics (PV). PV is the flagship of renewables. It provides clean power everywhere, on the grid or off, and in all sizes — from small cells that power pocket calculators to large solar power plants such as the 5-megawatt-peak plant that went online in the German state of Saxony in 2004. At the time, it was the largest photovoltaic power plant in the world.

Figure 8.1:
The Geosol plant on a waste heap in Saxony, Germany. (Source: Geosol)

So what is the world's largest PV power plant doing in Saxony, which has barely half the sunlight of the southwestern United States? Is Germany subsidizing a technology so much that it will pay for itself anywhere — a typical waste of tax money? Does it not make more sense, if we are going to have solar power at all, to put up giant solar power plants in the desert, such as in California, Nevada, and Arizona? Would it not make more sense for Europe to put up solar power plants in the Sahara or at least in Spain? Or should we just do away with photovoltaics altogether because it is nowhere near competitive?

Before we get to the answers to these questions, let us make a couple of distinctions. There are a number of applications that use solar energy but have nothing to do with electricity. For instance, solar cookers basically are just boxes that concentrate sunlight on a pot. And when you hang your clothes up to dry instead of using a dryer, you also are using solar energy. Indeed, photovoltaics is not even the only way to use solar energy to generate electricity. The largest solar power plant in the world is what is called a solar-thermal plant in the desert at Kramer Junction in California. There, solar heat is concentrated in parabolic collectors to heat a medium that flows through pipes. At the end of these pipes, the medium drives a conventional steam turbine. Kramer Junction consists of five such facilities, each producing 30 megawatts — a total of some 30 times more peak power than the photovoltaic power plant in Saxony.

Figure 8.2:
The solar-thermal power plant in Almería, Spain. At the top left and right, we see two wedge-shaped arrays of mirrors concentrating sunlight on a tower in the center. At the bottom right are the rows of parabolic collectors. Solar energy is concentrated on a medium — for instance, molten salt — stored at the top of the towers, while the medium in the parabolic collectors runs through pipes in each row. (Source: Spanish Ministry for Education and Science)

While the power at Kramer Junction costs around 12 cents per kilowatt-hour, a kilowatt-hour in Saxony will cost a full 41 cents, according to Gero Hollmann, managing director at Geosol, the company that operates the Saxony plant.[1] Nonetheless, the plant will be profitable because Germany's Renewable Energy Act stipulates that 45.7 cents must be paid for each kilowatt-hour of electricity from photovoltaics sold to the grid from a system that went live in 2004. In other words, the plant in Saxony has a profit margin of around 10 percent, making it an excellent investment if the company's figures prove to be correct.

One of the reasons the plant in Saxony is so profitable is that the land was almost free: the PV plant was constructed on a former waste site for a brown coal plant. A new project in the port town of Rostock on the Baltic Sea also takes advantage of cheap real estate. On a former waste dump at the port, solar tracking units have been installed. While these systems are more expensive than immobile PV systems, they also produce around 35 percent more power because they have an optimal angle to the sun almost all the time. One special feature of the 300 kilowatt-peak tracking units in Rostock is that they are not time-controlled but light-sensitive, which increases the output even more.

Günter Schmarje of Küstensolar, the company that operates the plant in Rostock, estimates that there are some 1,200 such waste dumps in the German state of Mecklenburg-Vorpommern alone that currently are not being used for any purpose and can be provided practically for free. He believes that, like the project in Saxony, the Rostock project will have a profit margin of about 10 percent.[2]

Is PV Too Expensive?

Of course, we could make just about any technology profitable by throwing enough subsidies at it. The retail price of a kilowatt-hour in Germany currently is around 19 cents; in North America, it is closer to 10 cents in most areas, though prices range widely. Roughly speaking, we can say that electricity is twice as expensive in Europe as it is in North America. And the 45.7 cents per kilowatt-hour of the Renewable Energy Act is still easily more than twice the retail rate in Europe. Would it not make more sense to develop solar-thermal power stations instead if they cost only a quarter as much?

The answer to this question shows us what is special about photovoltaics. While solar-thermal — and all other — power plants require their own real estate and have to be fairly large to be profitable at all (the new solar-thermal plant in Arizona is considered small at 1 megawatt), PV can be integrated into existing structures and tailored to local demand. In addition, solar-thermal plants are not expected to become drastically cheaper whereas photovoltaics is.

In Germany, Phönix SonnenStrom AG installed a 180-kilowatt system on a noise barrier along a train line.[3] No additional space is needed here and power is produced right along the power lines where consumption occurs, saving costs even further by reducing distribution losses. Indeed, power plants will not have to be expanded greatly in order to allow PV to provide a greater share of our power if we distribute the PV systems across rooftops. Instead, power will be produced right where it is consumed and the grid will increasingly be used only for emergency power.

Not only can PV systems be installed on roofs and façades but they also can *be* the roof or the façade. In such cases, the land on which the system is installed is already paid for and the money that would have been spent on the roof or façade is avoided, potentially lowering costs even further. For a normal house, the savings will be negligible because shingles are cheap, but quite expensive material is used on flashy buildings in modern business districts. For instance, the bronze façade for the new Westminster office complex in London cost around 7,000 pounds sterling per square meter. A PV façade would cost only a fraction of that — about $800 per square meter — and look no less impressive. One can argue that the power from such systems is free. This kind of thinking is one reason the CIS Tower in Manchester — the tallest tower in the United Kingdom outside London — received a new PV façade when the building was renovated. At the time, it was the largest vertical PV installation in Europe.

Architects are gradually catching on to this approach. In November 2003, the Energy Agency of the German state of North Rhine/Westfalia hosted a conference called "Power Replaces Marble." Architect Ingo Hagemann, author of the book *Gebäudeintegrierte Photovoltaik*[4] (Building-Integrated Photovoltaics), explained in an interview with me that it is very hard to provide any general figures about avoided costs. When the Zara fashion store was renovated in downtown Cologne, the PV façade replaced one that was much thicker, opening up more office space, which meant that the owner could receive more rent. And, Hagemann explained, building-integrated PV is an eyecatcher: "PV is sexy and allows architects to put up buildings that people are not yet used to seeing."

We also should not forget when discussing how much PV power costs that, unlike wind energy systems, photovoltaics produces the most power around noon, when demand is highest. Integrating a large share of PV power into the grid will thus be easier than integrating wind power, especially because PV will help "shave peaks." When demand for power is high, conventional plants can use all the help they can get, so more PV will mean that we will not have to build so many conventional plants.

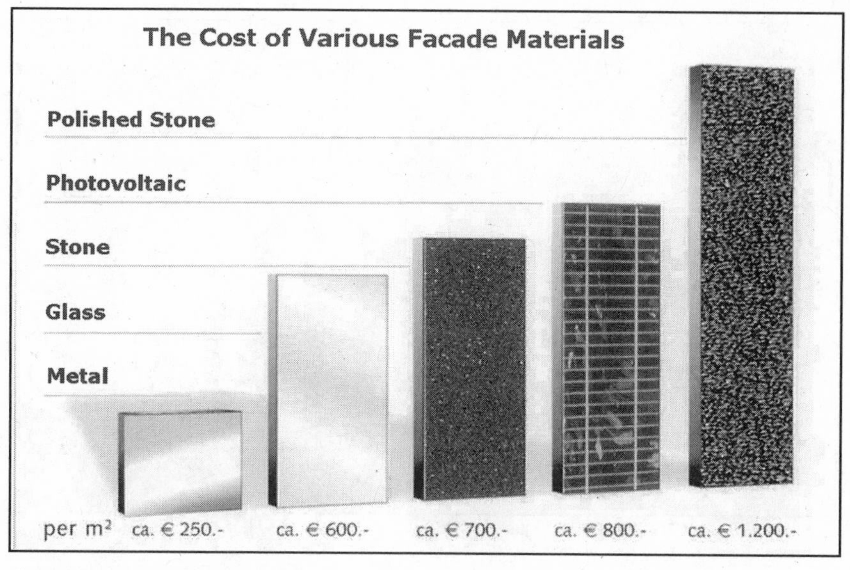

Figure 8.3:
Façade costs. PV often does not cost much more than other fancy façades. If the costs for these other materials are subtracted, the power from the PV system becomes very cheap indeed, if not free (when avoided costs of, say, polished stone exceed the costs of the PV system installed). (Source: Author's estimates)

The Potential of PV is Great

While theoretically the upper limit to the share of wind power in our overall power supply may be 25 percent (roughly the current level in Denmark and some parts of the grid in northern Germany), the upper limit for photovoltaics would be much greater because the supply so closely corresponds to demand. At the moment, however, photovoltaics is far from any theoretical upper limit. Even though it is booming in Germany at rates exceeding 30 percent per year, it still covered less than 0.1 percent of the country's power supply at the end of 2004.[5]

If we understand PV power as power to shave peaks, the price immediately becomes more competitive. On the spot markets in California and in Europe, the price of a kilowatt-hour of PV power has reached 50 cents — more than the base rate paid to producers of solar power in Germany. Since the Renewable Energy Act was revised in 2004, the basic rate for a kilowatt-hour of PV power in Germany has been 45.7 cents, but only for field systems like the one in Saxony. As discussed above, it makes a lot of sense to integrate PV into existing structures, which is why the Renewable Energy Act provided for payment in 2004 of 57.4 cents per kilowatt-hour up to 30 kilowatt-peak if the PV is integrated into buildings (including noise barriers). And if the systems are used on façades, 5 cents are added, resulting in payment for PV of between 45.7 and 62.4 cents per kilowatt-hour starting in 2004, depending on the type and size

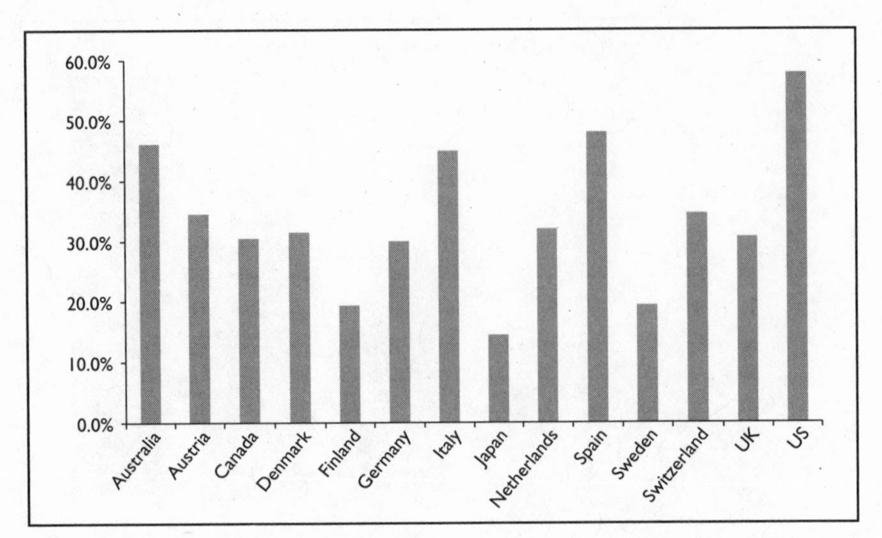

Figure 8.4:
The potential share of building-integrated PV in the total power supply of selected countries. According to the 2002 IEA study *Potential for Building-Integrated Photovoltaics,* building-integrated PV could cover between 14 and 58 percent of a country's current power supply. In Germany, which is just now overtaking Japan as the world's photovoltaics leader, the potential is just above 30 percent. This power would be generated at midday, when people need it most. Keep in mind that this potential includes only systems on roofs and façades, not field systems (solar power plants). Furthermore, off-grid applications are not taken into consideration here. In other words, the total potential of PV is much greater. The two technological leaders today — Japan and Germany — do not by any means have the greatest potential but they understand they will be able to sell modules to countries that do, such as Australia, the US, and Spain. (Source: IEA)

of the system. All of these rates are degressive — i.e., they drop slightly every two years for new installations but remain constant for 20 years at the rate specified for the year of installation. The idea is that economies of scale and technological advances should cause prices for the modules to drop.

At the moment, the exact opposite seems to be happening. The boom in Germany, coming at the end of a long boom in Japan, has put a strain on global supplies of solar silicon. German companies with full order books have been unable to produce at full capacity for lack of raw materials. This effect was felt all over the world, with companies in North America also complaining of steeply rising prices for solar silicon and solar cells. If the boom continues, we will have to find a way to produce large quantities of solar silicon inexpensively to ensure that the economies of scale will materialize. The technological advances fortunately already are weighing in, with companies reducing the thickness of solar cells so they require less solar silicon.

Germany's Renewable Energy Act has some of the highest rates for photovoltaics in the world. But it is not without competition: Luxembourg offers 60 cents for a kilowatt-hour and the government also covers part of the initial procurement cost, while Spain has instituted a similar program. But within Europe, not every country is toeing the line. The French program *bâtiments bleus*[6] (blue buildings) pays only 14.8 cents for a kilowatt-hour from a PV system. Fortunately, twice as much is paid in France's overseas *départements,* where imports of fossil fuels make electricity very expensive.

In the US, the main type of compensation is called "net metering." At the time of writing, almost 40 states in the US offered some type of net metering, with similar programs being offered in Canada. Here, the power meter basically runs backward when more power is produced than consumed. Owners of PV systems receive no more than the retail price of a kilowatt-hour of electricity (generally no more than 10 cents).

In some cases, owners of PV systems do not even get the full retail rate if they produce more than they consume. Rather, utilities sometimes pay only the avoided costs for this "excess" power — deducting costs for the grid and similar fixed costs — if they pay for excess power at all. No wonder Germany and Japan, whose programs to promote PV were similar, have overtaken the US, which used to be a world leader in PV but now has been relegated to third place with only about a fifth of the global market. Estimates are that Germany

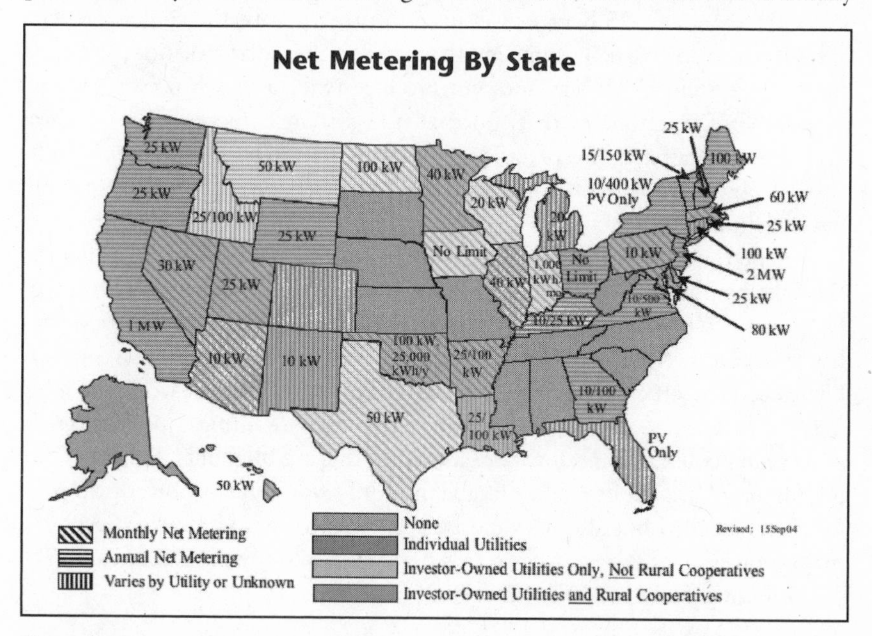

Figure 8.5:
A US Department of Energy map of net metering. The map unintentionally shows how chaotic state PV laws are. In Germany, one system of rates applies for the whole country. (Source: <www.eere.energy.gov/greenpower/markets/netmetering.shtml>)

now installs more than three times as much photovoltaics per year as the US does, slightly ahead of Japan.

Gray Skies?

Does it even make sense for cloudy Germany to try to be the world leader in solar energy? Yes, if we remember that sunlight may be free but solar power systems are not. Spain may have more sun but if Germany takes a head start and produces the best systems, it will be selling them to Spain. Anyway, Germany needs to include solar power because compared with other countries (such as Great Britain, France, and the US) it has not only little sun but also relatively little wind and even less petroleum and its coal resources are not competitive at world market prices.

In addition, people often fail to understand that photovoltaic systems usually run best at 25 degrees Celsius. A system in Mexico City would theoretically produce twice as much power (aside from considerations of smog) as a module in Stuttgart, which has half the solar radiation. However, a module in Mexico will quickly heat up to far above 25 degrees Celsius in the summer sun. According to Klaus Kiefer at the Fraunhofer Institute for Solar Energy Systems (ISE), the efficiency of silicon-based systems drops by about 0.4 percent for every degree above 25 degrees Celsius.[7] In other words, the cooler temperatures in Germany partially compensate for the weaker solar radiation, so that in practice a system in Mexico does not produce twice as much power. Overall, Fraunhofer ISE estimates that, once all the system components — such as inverters and batteries (for off-grid applications) — are taken into account, the cost of power from photovoltaics at the equator is only around 10 percent lower than in Germany.

Furthermore, it would not make sense to set up giant systems in Spain or in the Sahara and then have to transport the electricity to Germany just to benefit from a few percentage points of greater efficiency. Since no superconductor has yet been invented for such applications, the transport losses would probably exceed any efficiency gains. And even if one is invented, it would be a big mistake to become dependent on foreign electricity. During a political crisis, power lines could be shut down or sabotaged, causing blackouts. If every house has photovoltaics on its roof or façade, not only would many power pylons no longer be needed but also it would be very hard to cause a blackout even if a major power plant is destroyed or fails. So if cloudy Germany invests in photovoltaics now, the future looks sunny and bright.

The Cost of Photovoltaics

Around 98 percent of photovoltaics modules consist of silicon. One often reads that silicon is the most common element in the Earth's crust and the second most common element on the Earth (after oxygen). Unfortunately, such statements are misleading because silicon does not exist in a form that can be used for the production of industrial solar cells. It is found mostly in combination with oxygen in quartz sand. (Similarly, hydrogen, which also is misleadingly touted as being plentiful, does not naturally occur in a form that can be used in fuel cells.)

Is there any way to generate solar power from photovoltaics without using silicon? How cheap can photovoltaics become? In which applications is photovoltaics already competitive despite the current high price? To begin with, there are three types of photovoltaic cells: monocrystalline, polycrystalline, and amorphous (also called thin cells). Mono cells have the greatest efficiency at around 16 percent, with poly cells being only slightly lower at 15 percent. However, mono cells are more expensive because their silicon has to be of greater purity. Single crystals are drawn out of fluid silicon to make mono cells, while poly cells are made when wafers are sawed out of solidified blocks of molten silicon. When this molten silicon solidifies, many crystals are formed. The fissures between them reduce the efficiency of the cell somewhat, but the manufacturing process is less expensive.

Up to now, solar silicon has mainly been produced from the waste products from the manufacture of semiconductors because semiconductor-grade silicon is of even higher purity than solar-grade silicon. In discussions about the environmental impact of photovoltaics modules, people often ask about the agents used in the manufacture of such solar cells, but two issues are usually overlooked. First, solar cells are already recycled products from the semiconductor industry and, second, the solar silicon used in photovoltaics modules can itself be reused when the modules are replaced after 30 years or longer. The silicon does not age; only the glass plates and lamination that protect the cells from the elements do.

The situation is a bit different with amorphous cells, though recently some progress has been made toward recycling such cells. To make these cells, a thin layer of silicon is vaporized on an inexpensive carrier such as glass. The process of removing this thin layer of silicon has been an obstacle to making its recycling economical. However, so little silicon is used that little is lost if it is not recycled. As the cells make do with so little silicon, they are the cheapest of the three main types of solar cells. On the other hand, they have by far the lowest efficiency at around 8 percent.

We now see why the mere indication of the efficiency of a solar cell does not tell us much. Not only do we have to ask how much power we get for our money but we also have to take into account how much energy the cells produce under given conditions. Under diffuse light conditions, thin cells may produce as much energy as polycrystalline and monocrystalline cells. In other words, if the sky is cloudy or the photovoltaic system has a less than optimal orientation, inexpensive thin cells actually may produce as much power as crystalline cells. Thin cells therefore often are used on façades at a 90-degree angle to the ground. At the border between Canada and the US the optimal angle would be closer to 30 degrees, approaching zero as we near the equator. If we wanted to clad a skyscraper's or an office building's façade with photovoltaics, it might make sense to apply thin cells to the eastern and western sides and crystalline cells to the south side.

Lower Prices from Mass Production?

The question of what kind of cell we want to use therefore has to be decided on a case-by-case basis. But there is no denying that power from photovoltaics is not cheap. Since 1990, prices have dropped by around two-thirds from approximately $15,000 per kilowatt-peak to less than $6,000 per kilowatt-peak in 2005. Experts have been promising that prices will continue to plummet as mass production ramps up, just as happened in the computer industry. As Helmut Engel of Sharp Deutschland put it in 2003, "In four or five years, solar cells will be economical, and a long growth phase will ensue." [8] In February 2004, a spokesperson from Apex-BP Solar was quoted in *Le Monde diplomatique* as saying that the price of a kilowatt-hour of solar power will reach the level of conventional energy carriers in Japan and California by 2010.[9]

Don't hold your breath. The boom in Germany has shown that there is not enough waste silicon from the semiconductor industry to go around. The photovoltaics industry is now looking for ways to produce solar silicon cheaply on its own. Nonetheless, the EU expects that the price per kilowatt-peak will drop to 3,200 euros by 2030. By then, a kilowatt-hour of power from photovoltaics would still cost 17 cents. As the EU report put it, photovoltaics would still be "generally uncompetitive." [10]

Researchers at Fraunhofer ISE announced at the beginning of 2004 that they had developed a silicon cell only 38 micrometers thick with an efficiency exceeding 20 percent. [11] But even as manufacturers feverishly develop ways to manufacture ever-thinner silicon wafers to lower material costs, some industry observers are hoping for a breakthrough outside of silicon. In 2000, Alan Heeger won a Nobel Prize for developing a conductive plastic that could lead the way to the invention of cheaper solar cells. [12] In April 2004, the US company TDA Research and the National Science Foundation announced the development of a new conductive polymer called oligotron. [13] Researchers at the Lawrence Berkeley National Laboratory in the US have developed a metal alloy that could raise the efficiency of solar cells to 45 percent or more. [14] Finally, there are Grätzel cells, which use pigments to reproduce the function of chlorophyll in photosynthesis to generate power. [15] The main obstacle here is longevity as the efficiency of these cells begins to drop almost immediately.

At the moment, such applications are pie in the sky. No one can say for sure when photovoltaics will be as cheap as power from coal plants. But the trend is clear: every year photovoltaics gets about 4 percent cheaper and our coal resources become scarcer.

Off-Grid Applications

In some applications, photovoltaics is already the cheapest source of power despite its high price — even without any subsidies. If your home is not close to the grid or your local utility company will not pay you enough for your power, photovoltaics generally is cheaper than hooking up your house to the grid. Diesel generators usually are not sold with outputs below 5 kilowatts and even if you have one it is uneconomical to leave it running all the time to supply a little bit of power. In contrast, photovoltaics provides power all day.

Globally, around half of the installed photovoltaics capacity is off the grid. Applications include computer keyboards, portable radios, parking meters, and telematics systems along highways. In small devices, it pays to replace large batteries with small rechargeable batteries, while connections to the grid would often be more expensive even in the case of parking meters and highway cameras. Even in the neon lights of Las Vegas, some bus stops have lighting powered by photovoltaics because it would be too expensive to rip up the streets and sidewalks so that a power line could provide the little power needed. At bus stops, the solar cells charge a battery during the day that then powers a lamp at night.

Such a power supply represents a real breakthrough for mobility. For instance, it is now commonplace for workers to take their office with them in a laptop and cell phone, but at some point you need to recharge. Recently, solar power systems have been developed that allow you to take your laptop and cell phone all the way to your remote mountain cottage to work in peace and quiet.

Of course, your cell phone does not have reception in your mountain cottage, but just wait. One of the major obstacles to expanding mobile communications networks is providing power to remote base stations. It turns out that photovoltaics systems that charge batteries are the least expensive source of power for many such applications. In many remote parts of the world, cell phone networks are being expanded thanks to photovoltaics. And that is not all: in Japan, a new solar charger is being sold to provide power for cell phones after an earthquake.

Photovoltaics also is used in what are called microgrids or minigrids. Here, a small hybrid electricity grid is set up for a village. Such microgrids often start off with a diesel generator that charges batteries. Photovoltaics systems are added and power lines are installed between houses to create a small power grid, with a diesel generator providing only emergency power. Other combinations are also used, such as a wind turbine with photovoltaics. Here, a diesel generator rarely needs to be used because wind energy and solar energy complement each other so well. In the Philippine village of Atulayan, Synergy Power set up such a microgrid to replace the old power system that ran on a diesel generator. According to the company, residents of the village used to have only four hours of power per day. They now have power around the clock. The diesel generator switches on for only a few hours every few days when the batteries' charge drops too low.[16]

Central Power Plants as Emergency Power Generators

Once a microgrid or a solar home system (SHS) — the latter generally consisting of a solar panel, an inverter, and a battery to power energy-efficient lamps, a radio, a black-and-white television, and such — has been installed, demand for power generally increases and the system can be overburdened. Users may become frustrated and believe that such systems do not work. Of course, it is quite easy to expand these systems, but that costs money. Diesel generators are cheaper upfront and fuel costs keep consumption down. With photovoltaics, you pay for everything upfront.

How are poor people to come up with the money? In addition to state subsidies and development aid, there is the option of microfinancing, such as from Grameen Shakti (part of the Grameen Bank network). In such programs, small sums of money are loaned, generally to women who want to increase their income by purchasing, for instance, a sewing machine and a lamp. In such cases, buying an SHS increases the income of the user but in other cases — such as when power is provided to light schools and community buildings — the connection between a higher standard of living and a higher income may not be obvious. Sometimes the income of recipients of micrograms for SHSs does not rise as quickly as their standard of living and they cannot afford to expand the system themselves.

Even in densely populated, wealthy Europe, where the grid reaches almost everyone, hybrid grids are used in remote areas such as the famous Rappeneck hikers' cottage in the Black Forest, where a fuel cell recently was added to a system consisting of a wind turbine and photovoltaics. A similar wind/photovoltaics/diesel hybrid system has been in operation since 1996 at a remote cottage in Bavaria called Rotwandhaus. Tiny, remote villages in the mountainous areas of Spain and Greece also use such systems.

Some two billion people — a third of humanity — do not have access to the power grid. These people spend a much larger share of their income on energy than we do. Despite their poverty, they spend up to $10 a month — a third of their income in some cases — on fuel, batteries, candles, and such.

Hybrid systems consisting of photovoltaics and wind energy offer far more energy for less money and these renewables are much cleaner and healthier than the sooty oil lamps that still are considered luxury items in the poorest countries. The main problem is attitudes: people have to learn how to use the new technologies. Everyone knows that they simply switch on a diesel generator when they need power or when batteries are low, but when there is no wind and sunlight, you cannot just buy more fuel. You have to plan ahead.

It is not only the microgrids of developing nations that require some rethinking. For example, the network operators in Denmark and Germany generally have to buy the power that wind turbines produce whether they want it or not. To keep the grids stable, they may have to lower the output of large, central power plants. In other words, nuclear and coal plants are increasingly performing the functions of a diesel generator in a hybrid microgrid, the main difference being that a diesel generator is generally not run all the time and provides only a fraction of the total power, whereas central power plants cover almost all our power needs and are switched off completely only for maintenance and repairs.

Not only are operators of central plants going to be losing shares of the market but their plants also will not be running at capacity whenever the wind blows or the sun shines. While forecasts of wind have reached an impressive level of accuracy in Germany so that surprises are rarely in store, we still do not have wind when we want it. From the point of view of utilities, central power plants are being "degraded" to the level of emergency power generators.

Central power plant operators have been claiming that, ironically, even if we add a lot of wind turbines, new power plants will have to be built so that the central plants can still provide full power when there is no wind. This argument is fallacious. Since the wind is always blowing somewhere on the grid, wind energy is never completely lost. What the utilities are really complaining about is that their plants are having to run below full capacity, and large coal and nuclear plants are not economical when they do not run full blast to cover the base load.

Energy Payback

Once and for all we must dismiss the claim that photovoltaics does not produce as much energy as is invested in it during its manufacture. This claim not only is wrong; it also underscores the great advantage of photovoltaics.

Let us start with the seemingly true statement that a coal plant produces a lot of power. For example, 600 megawatts sounds like a lot — you could run 600,000 toasters at one time with such a plant. Let us now assume that this plant has an efficiency of 33 percent: the plant converts one-third of the energy stored in the coal into electricity and two-thirds into waste heat. While it is true that this coal plant will convert far more energy into electricity than was required for its construction, it also is undeniable that we will put far more energy into such a plant than we will ever get out of it. Indeed, we put three times more energy into such a plant than we get out of it. One day, when we have run out of coal, we will be forced to realize this.

In contrast, we will never put more energy into photovoltaics than we get out because we are not exhausting the sun by generating electricity with solar cells. The sun will be around for a long time and will probably vaporize the Earth before it destroys itself in another billion years or so. Until then, we should use the energy that the sun sends us every day — 16,000 times our global energy consumption. Every year, Germany gets around 1,000 kilowatt-hours of solar energy per square meter (roughly ten square feet). That is about the amount of electricity the average German consumes. At an efficiency of 12.5 percent, that average German needs only eight square meters of photovoltaics to cover all electricity needs. North Americans consume about twice as much electricity, but more efficient consumption could lower this amount and solar radiation is greater in most parts of the US than in Germany. Americans probably would need less than 150 square feet of PV per person to provide enough electricity to cover their needs.

Even ExxonMobil admits that a photovoltaics system produces as much energy in two years as was invested in it.[17] Proponents of photovoltaics, such as Germany's Irm Pontenagel, have come up with more conservative estimates for the energy payback of photovoltaics by including inverters, frames, and such. Pontenagel claims that the energy payback of a photovoltaics system is about three to six years. In that case, a photovoltaics system produces some five to ten times the energy we invest in it. Compare that to constantly losing two-thirds of the energy in a coal plant.

The efficiency of a coal plant therefore cannot be compared to the efficiency of renewables. A coal plant with 33 percent efficiency is by no means twice as efficient as a monocrystalline cell with at least 16 percent efficiency. The sun sends us far more energy than we could ever take. Let's start using more of it.

Chapter 9

Wind

Opinion surveys regularly show that just over eight out of ten people (80%) are in favor of wind energy, and less than one in ten (around 5%) are against it. The rest are undecided. Public opinion in support of wind power tends to become even more strongly in favor once the wind turbines are installed and operating, a finding from several surveys carried out in the UK and in Spain.

—The American Wind Energy Association, 2005

Germany is the largest producer of wind energy in the world, although France, Great Britain, and Spain have far greater potential. Of all countries, the US has the most wind energy potential by far. Will any technology work if you throw enough money at it? That is the way many critics see it, among them Germany's premier weekly news magazine *Der Spiegel*, which in March 2004 published a critique of Germany's wind energy success.[1] The critique demonstrates the double standards applied to wind and conventional energy in both Europe and North America.

The complaints are well-known: wind energy is not competitive without subsidies; it is not truly environmentally friendly because coal and nuclear plants have to be built anyway to fill in the gaps when the wind is not blowing; and wind turbines ruin landscapes. One person was even quoted in *Der Spiegel* as saying: "Wind farms have destroyed German landscapes worse than anything since the Thirty Years' War." In this war (1618-1648), Germany was devastated. Incredibly, in addition to World War II, in which German cities were bombed into rubble, this critic left out a lot of other things that have changed the German landscape considerably. These include industrialization, in which factories and power plants polluted the air and poisoned entire bodies of water (the Rhine, Germany's main river, was completely dead only a few decades ago after a chemicals accident upstream in Switzerland); roads and train tracks; urban sprawl (which admittedly is not as bad in Europe as in North America);

Figure 9.1:
If not wind, then what? See <www.ifnotwind.org> for further discussion. (Source: *Das Solarbuch* by Dieter Seifried and Walter Witzel)

and, last but not least, the approximately 180,000 power pylons. By comparison, Germany had only around 17,000 wind turbines in 2005.

So why is everyone getting upset? Actually, few people are. *Der Spiegel* claims that the resistance to wind power is growing, but polls have repeatedly shown that wind power remains by far the most popular source of energy not only in Germany but all over the EU and in North America. Nonetheless, in the *Der Spiegel* article, which also made headlines in Britain, German poet Botho Strauss is quoted as saying: "No other phase of industrialization has caused a more brutal distruction of landscapes and we have done propping up wind turbines all over the place."[2]

Such a statement is hard to quantify. The article mentions "400,000 migratory birds" that *could* be slaughtered each year by the "bird blenders" in a planned wind farm off the coast of Germany. However, many studies have come to other conclusions, such as one to three birds being killed per turbine per year. On the other hand, artificial reefs are created underwater beneath offshore turbines, creating a new habitat for certain species of fish and mussels.

Whatever the actual figures, it is clear that a double standard is being applied here. Wind power has to fulfill requirements never placed on coal or nuclear power plants, and when the wind power industry receives subsidies on the level of what coal and nuclear power have long been receiving, people complain about the subsidies for wind.

It is quite probable that wind energy actually saves the lives of more birds than it kills. While various studies have found that a typical wind turbine kills a few birds every year, when wind power replaces coal power the air becomes cleaner, benefiting plants, animals, and humans. We can entertain the argument against wind power only if we act as though we had a choice between pristine landscapes and wind power, or — as *Der Spiegel* put it — "your money or your landscape."

Why were these standards never applied to coal power? The proper comparison, assuming that we do not want to do without electricity altogether, would be between wind power and other types of energy such as coal. But then the comparison becomes difficult because no one has ever done a scientific study to determine how many birds die from the emissions from a coal plant, aside from the rough estimate of a Canadian engineer who found that the number of birds saved by one wind turbine in his project amounted to 1,710 per year.[3]

NIMBY

We should not be surprised to hear that there is sometimes local resistance to wind farms. After all, people are generally against everything that looks like industry as

Figure 9.2:
Now, be honest. What do you find unattractive in this picture? (Photo courtesy of the German Wind Energy Association)

soon as there are plans to put it up in their backyard. Indeed, in recent decades a clear trend toward racism has become apparent in the United States, with industrial plants being erected in areas predominantly populated by minorities.

It would seem that we all want clean power but we do not want to see anything except the power socket. Most Europeans probably would like to have a giant solar power plant in the Sahara feeding power to Europe via a superconductor that has yet to be invented — and of course without any troublesome political turmoil that might detrimentally affect "our" power supply. Did I forget to mention that the power should be cheap? The coal industry has devoured entire villages and flattened mountains. Wind turbines can be dismantled at any time without leaving a trace. They do not change landscapes permanently.

The question of subsidies is another double standard. *Der Spiegel* estimates that each of the estimated 40,000 jobs in the wind industry at the end of 2003 received 21,750 euros in subsidies. Assuming that figure is accurate, it is a fraction of the subsidies that coal receives in Germany: an estimated 80,000 euros per job.[4] Indeed, the subsidies for all renewables in Germany make up only 22 percent of the total subsidies for the coal industry. At the same time, when the article was published at the beginning of 2004 a total of 130,000 people were working in all the renewables sectors, compared to 90,000 in the coal sector.[5] Jobs in the coal industry are being cut, while the renewables sectors are booming. Probably the *Der Spiegel* estimate of the number of jobs in the wind industry was too low to begin with: the daily *Süddeutsche Zeitung* of Munich put the number at the end of 2003 at 45,000.[6]

Proponents of coal claim it is unfair to list the amount of subsidies per job in the coal industry. They prefer to point out that, since coal produces seven times as much energy in Germany as renewables do, coal subsidies are lower per kilowatt-hour. But there is a flip side to this coin: renewables are more expensive but they also create more jobs — exactly what a weak labor market like Germany's needs. This money circulates within the country. Instead of importing energy, a country can have a strong energy export industry.

This argument has been maliciously misunderstood in Germany. One of my critics claimed that, according to my thinking, we might as well pass a law making it mandatory for video recorders sold in Germany to be hand-carved out of wood to boost the domestic hand-carving industry. What this person overlooked is that there will never be a world market for hand-carved video recorders but there will be one for renewables. The German renewables industry and German politicians are banking on great world demand for renewables as fossil fuels become scarce. Germany is setting up an industry in anticipation of foreigners standing in line for German equipment in the future when humanity has no other choice. According to Stephan Kohler, executive director of the German Energy Agency (dena), exports of renewable energy technologies probably will create more than 200,000 additional jobs in Germany by 2020.[7]

Shadow Power Plants

But the double standards do not end there. Because wind power is booming, the grid has to be expanded in some areas. Critics then complain as though the grid did not have to be expanded every time a central coal or nuclear plant goes online. Ironically, the grid does not have to be expanded nearly as much for renewables as for central power plants because most renewables generate distributed power, close to the consumer.

What about the claim that wind energy does not actually reduce emissions that much because large power plants waste a lot of energy when they are ramped up and down to accommodate fluctuations in wind power? This argument is simply disingenuous. What utility companies are really complaining about is that they are losing market share and are not able to run their plants at full capacity as often, which cuts into their margins. At the beginning of 2004, the "Big Four" utilities in Germany charged that they would have to increase their prices because the growing share of renewables was cutting into their margins. They invented the term "shadow power plant" to describe how their power plants had to be "on standby" in case the wind stopped blowing.

Unfortunately for them, it turned out that the grid with the most power on reserve was RWE's — the grid that also had the smallest share of wind power. In addition, it seems that utilities were charging far more than the extra costs attributable to renewables. The figures given in *Der Spiegel* for the costs of these shadow power plants was 100 million euros. The German Association of Renewable Energy pointed out that utilities actually charged their customers an extra 500 million euros in 2003, allegedly for the added costs from renewables.[8] The dispute about the exact figures finally led to the creation of Germany's regulatory authority for the electricity grid in July 2005.

It is important to remember that utilities do not have to build new power plants even if we put up giant wind farms, as they sometimes claim, since there will always be a certain amount of wind across an entire grid. But there is no denying that sharing the grid with distributed producers of renewable power cuts into the margins of utilities. Coal and nuclear plants also require stand-by capacity when they go offline for maintenance and repairs.

In the end, the *Der Spiegel* diatribe against wind energy only showed that the critics were not going to change their arguments no matter what. In an article from 1997, the magazine had already complained because wind power made up only 0.7 percent of total German electricity generation.[9] In 2004, the magazine complained that it was only 4 percent.

Let us take a closer look now at some of this criticism that never dies.

Bird Blenders?

Wind turbines are bird blenders, and wind energy itself way too expensive — at least that's the way some critics would have us view the technology that is now the most successful of all modern types of renewable energy. For more than 20 years, the wind power industry has been reacting to such accusations and improving technology. The result: the criticism hasn't changed and some comparisons are eye-opening.

The oil crises of 1973 and 1979 revealed the critical dependence of the US economy on oil. President Carter responded by implementing the nation's first wind and solar programs. One of the largest wind energy projects began in 1982 at Altamont Pass east of San Francisco. By 1987, a total of 7,340 turbines had been installed — right where a number of endangered raptors flew. In the 1960s, the population of one of the species, the bald eagle, had already been decimated to around 30 birds because of the thinning of their eggshells from the effects of DDT. Without DDT, we may never have thought of wind turbines as dangerous to birds.

The first study of the effects of the wind turbines in Altamont was published in 1992.[10] Of the total of 7,340 turbines, 1,169 were studied for a period of two years (1989-1991), but the study was limited to endangered raptors, with other species only being recorded haphazardly. The study found that these 1,169 turbines had killed 182 raptors — one bird per 13 turbines per year, or 0.07 per turbine per year. However, the number of all birds killed by the rotor blades is much higher. A 2003 NREL study put the figure at 0.19 birds per turbine.[11] A study published by the US National Wind Coordinating Committee (NWCC) in August 2001 estimated that the roughly 15,000 wind turbines then operating in the US killed around 33,000 birds annually,[12] some 2.2 birds per turbine per year, though this figure fluctuates greatly from one location to another.

How Many Are Too Many?

The NWCC report emphasizes that even if a million wind turbines were installed, the number of birds that would be killed by the blades — an estimated 2.2 million per year — would still be only a fraction of the damage that other man-made structures already cause. Buildings (windows) kill an estimated 500 million birds in the US every year (one researcher estimated the figure recently at 1 million and called for lights to be turned off in skyscrapers at night), cars and trucks 70 million, and telecommunication antennas about 27 million.

The estimated maximum of 2.2 million birds that may be killed per year in the US by wind turbines in the distant future pales in comparison, not to mention the estimated 33,000 for 2001. But has anyone ever called cars — or buildings, for that matter — bird blenders? It should be noted that bird protection groups took the Federal Communications Commission to court in 2003 because environmental impact assessments (EIAs) were not even mandatory for many communications facilities.

In contrast, EIAs have long been required for wind turbines, despite the drastically lower danger they pose for birds. The 2002 NWCC manual for wind turbine permits pushes the comparison even further when it points out that the Audubon Society estimates that cats (wild and domestic) kill some 100 million birds each year in the US.[13] Indeed, the Audubon Society — the largest bird protection group in the US — is hardly an enemy of wind energy as some, such as a speaker from the right-wing Cato Institute on the Diane Rehm show in November 2001,[14] would have us believe. In June 2001, five months before the broadcast, Audubon's spokesperson John Bianchi answered my query about his organization's position on wind power as follows:

> At Audubon, we believe wind power is a great, non-polluting alternative to fossil fuels. We have only one reservation: wind generation plants must be located away from habitats for endangered birds, especially raptors, which have a higher chance of impacting with wind turbines. With the proper EIS work, wind plants should be a great benefit to people and the environment.[15]

The same position is held by environmental groups worldwide. In addition to Greenpeace, for instance, Germany's nature conservation group BUND has published a position paper on wind energy in which it clearly states:

> BUND welcomes the further expansion of the use of wind energy in Germany as a distributed source of renewable energy. This expansion has to take into account its impact on nature and mankind. As an especially environmentally friendly, sustainable source of energy, wind power will play an important role in the power supply from an ecological mix of energy sources towards a sustainable energy sector.[16]

This positive attitude is based on the understanding that wind power does not cause any air or soil pollution (acid rain), which affects birds directly and severely. Indeed, once the number of birds whose lives are saved by wind turbines is entered into the calculation, the results are incredibly positive. As mentioned above, the study for a new wind farm in Ontario estimated that the wind power that would replace the power from coal-fired plants would reduce pollution so much that 1,710 birds per turbine would be saved annually. Take away the 2.2 birds killed by the rotors and the balance for wind energy, +1,707.8 birds per turbine per year, is not bad. No wonder bird protection groups are in favor of wind power.

From the US to Europe

Wind farms in the US thus do not seem to have a significant negative effect on bird populations. Since Altamont, bird studies have been conducted in over a dozen states ranging from Tennessee to Minnesota and not one has shown that bird populations have been affected. Nor does wind power pose a threat to birds in Europe. The Danish Wind Industry Association (whose website offers an exhaustive history and current overview of wind power) puts it this way:

> Birds often collide with high voltage overhead lines, masts, poles, and windows of buildings. They are also killed by cars in ... traffic. Birds are seldom bothered by wind turbines, however. Radar studies from Tjaereborg in the western part of Denmark, where a 2 megawatt wind turbine with 60 metre rotor diameter is installed, show that birds — by day or night — tend to change their flight route some 100-200 metres before the turbine and pass above the turbine at a safe distance.[17]

A German dissertation came to similar findings for that country: "The effects of wind turbines on small birds observed during the day at four locations are estimated to be low. The danger of birds being killed by rotating rotor blades was also found to be low at the five turbines studied."[18] And in the offshore field, which will be booming in the next few years in Europe, no danger is expected to emanate from the turbines. In the Netherlands, mussels will even be farmed under them.

Learning from Mistakes

As one would expect with a large pilot project, researchers learned a lot from the mistakes made at Altamont. First, environmental impact assessments were made mandatory to see whether any animals, especially birds, would be affected. Then the design of wind turbines was changed to make them less dangerous for birds. In particular, the speed of the rotor blades was reduced.

Figure 9.3:
The world's first large wind farm at Altamont, California (east of San Francisco). Today a single turbine could replace all those shown here. Modern wind farms in the US — such as the one in Lamar, Colorado, near Denver — are much better spaced with less visual impact. The wind industry has learned many lessons from Altamont. (Source: Photo by the author in January 2006)

This slower speed did not reduce power generation. On the contrary, the first turbines at Altamont had a capacity of 55 kilowatts. Now, 2-megawatt turbines (that's 2,000 kilowatts) are on the market and the first 5-megawatt turbines are being developed for offshore use. Not bad: an increase of over 18,000 percent in about 25 years. At the same time, the price of a kilowatt-hour of wind energy has fallen from over 30 cents to around 5 cents in good locations. Wind power is thus cheaper than nuclear power and can compete with coal and oil — even if we do not include the great external costs for these energy carriers. Indeed, wind energy has become so cheap so fast that Germany — the world's wind leader in terms of installed capacity — revised its Energy Feed-In Act in 2004 to reduce the rates that owners of new wind turbines get for power they sell to the grid. (The rates are guaranteed for 20 years once turbines are on the grid.) Wind energy just doesn't need as much support anymore. Wind power is quite a success story. And, my fellow Americans, it's more a European success story than an American one.

A European Success Story

There remain two common bones of contention: wind turbines are said to be 1) loud and 2) ugly. Ever stood under a modern wind turbine turning full-blast? You may not be able to hear it if the leaves in nearby trees are rustling, as they will be when the wind is blowing. Several advances around 1990 made wind turbines quieter. First, as we saw, the rotor blades were slowed down, for the tips of the rotor blades are one of the main sources of noise. But the major advance came when rotor blades were developed with adjustable pitch. The blades now can be turned into and out of the wind, allowing for more optimal rotor speeds at various wind velocities.

Another major breakthrough came in 1992 from Germany. Enercon developed a gearless "direct drive" wind turbine, the E-40, that was more robust, powerful, and quiet than anything before.[19] Noise and energy losses from gears were now a thing of the past. You can't hear a modern wind turbine near a street over the noise from the traffic — and that has been the case for many years, as the website of the Danish Wind Industry Association explains:

> A survey on research and development priorities of Danish wind turbine manufacturers conducted in 1995, however, showed that no manufacturer considered mechanical noise as a problem any longer, and therefore no further research in the area was considered necessary. The reason was that within three years noise emissions had dropped to half their previous level due to better engineering practices.[20]

Some people still claim to be disturbed by an inaudibly deep droning said to emanate from wind turbines. The flickering shade caused by the rotating blades is another reason wind turbines should not be erected where they would cast shadows on buildings. No problem: just put them a few hundred yards away.

But modern wind turbines are so quiet that densely populated countries such as the Netherlands — where onshore space for wind turbines is dwindling — are looking for new ways of integrating small wind turbines into architecture.

Don't expect to see any 60-meter-tall wind turbines in the *grachten* (canals) of Amsterdam, but don't be surprised either if you soon see smaller models popping up on the roofs of the town — and producing more electricity over the year than the residents of the houses consume. Indeed, discussions about such building-integrated systems are by no means limited to the Netherlands. Researchers from the University of Stuttgart in Germany and the British Rutherford Appleton Laboratory are also designing buildings that concentrate wind for the turbines.[21] The first experiment models have already been built.

At the same time, wind expert Paul Gipe warns of such developments. Small wind generators are popular in North America, where they complement solar power units well in off-grid applications. But as Gipe points out, homes are

Figure 9.4:
This building is just a vision, but the Dutch are looking for alternatives to generate wind in the little space they have. What will the noise levels be inside the building and in the surrounding area? What kinds of stress will the turbine pass on to the building? Will the turbine generate enough power to pay for the extra costs of installing it inside the building? The questions remain unanswered, but the concept shows that wind turbines have become mature and quiet enough for such options to be considered. (Source: Sander Mertens, Technical University of Delft, Netherlands)

rarely built with the addition of wind turbines in mind. Such units can damage the architecture of a building or simply cause noise during operation as vibrations are carried through the house. Furthermore, Gipe correctly argues that it generally makes more sense to invest in wind farms in optimal locations as so many such sites remain unused.

ESTHETIC AND POLITICAL DECISIONS

The decision about whether wind power will make its way into urban centers will depend largely on how people react to the idea. As the saying goes, there is no accounting for taste, and it appears that many people do not find wind turbines unattractive. For example, in Australia, the turbine on the island of Rottnest could well become a tourist attraction. A recent study in North America found that wind turbines do not impact residential property values negatively (unlike high-voltage power pylons), and 80% of Europeans surveyed do not have a NIMBY attitude to planned wind farms but openly welcome them.[22]

When deciding how close to live to wind farms, we should not be distracted by arguments that these towers are not part of the traditional landscape or national heritage, as is often argued by detractors of wind power in Europe. Not only is this line of thinking quite rare even where it is advocated most vehemently — such as in Germany's Black Forest, where five of seven people recently surveyed stated that they find wind turbines attractive — but it also is generally nonsensical — such as in the Black Forest, where roads, train tracks, and ski slopes cut into the forest over the past 100 years have altered the landscape considerably. Acid rain from fossil-fuel pollution damaged the remaining forest so much that the term *Waldsterben* (forest death) was coined a few decades ago. And that is exactly the point. This "protect our heritage" line of thinking obscures the fact that the choice is not between pristine landscapes and monstrous wind towers but between coal, crude, and Chernobyl on the one hand and clean energy and eternal energy independence on the other. The only other option is an energy shortage.

EUROPE LEARNS FROM AMERICA'S MISTAKES

In the 1980s, the US had over half of the installed wind capacity worldwide. The US gradually lost its leadership when the cuts Ronald Reagan made to Jimmy Carter's renewable energy programs began to undo the initial progress. Throughout the 1990s, continuing uncertainties in the government's commitments to renewable energy made investments in this budding industry a bit of a rollercoaster ride. One year, federal support is good but the next year hardly anyone is willing to invest as governmental support is reconsidered. Wind turbines run for decades, so a stable investment plan has to be based on price commitments that last longer than a few years.

The US Energy Policy Act of 2005 continues support for wind power for two years. Compare that to support provided for the hydrocarbons industry, some of which is guaranteed in the Act for 20 years. Price commitments for wind are why Denmark and Germany have gained so much ground since the early 1990s. The political commitment to wind power in Germany, for instance, crosses all parties: from 1998 to 2005, the government was from the left (Social Democrats and Greens), but the government that first implemented price guarantees for wind power producers in 1990 was the right-of-center coalition under Helmut Kohl. Fifteen years later, this support is still in effect. Unlike their American counterparts, German investors do not have to include governmental wavering in their risk assessment. In 2003, three-fourths of the installed wind power capacity worldwide is found in Europe.

The US has recently had trouble competing with Europe in wind energy, which may be why there were reports in the German press (but interestingly not in the US press) of the US using the "Echelon" economic espionage system to steal wind technology from Germany. Enercon's E-40, the best selling turbine of all time, could not be sold in the US because an American firm — the now bankrupt Kenetech Windpower, Inc., which was subsequently taken over by GE, the leading wind turbine manufacturer in the US — filed a patent identical to the design of the E-40 before the Germans could bring their patent to the US. When the CEO of the German company was taken to court in the US, the Americans even presented photographs of the inside of an Enercon turbine. Apparently, they had cracked the code to enter the structure via Echelon and taken photographs, as reported in the German press and communicated to me by witnesses.[23]

Remember though that Germany may now be the number one producer of wind power but it hardly has good wind conditions. France and Great Britain, for instance, have much more potential but less political support for wind power. Overall, Germany produced 18 percent more "green" power in 2002 than in the previous year, and while this growth rate has dropped somewhat in recent years it still remains in the two-digit percentage range. So while the wind may blow and the sun may shine for free, the technology to harvest this power costs good money. The question is: How much of it will be "Made in the USA"?

Chapter 10

Geothermal Power

Economically, there is no alternative to using renewables more.
—Claudia Kemfert of the German Institute for Economic Research, 2005

For the past 150 years, geologists have surveyed the Earth looking for "black gold." Now the same drilling technology is being used to tap a new source of energy: geothermal power. Unocal, a well-known American oil and gas company, is also a major geothermal producer, for instance in the Philippines, which already gets more than a fourth of its power from geothermal plants and may become the global leader in geothermal power in the next few decades. The first geothermal power plant went into operation 100 years ago in Larderello, Italy. But, as with most renewables, geothermal really got started only at the end of the 1970s in the US. Currently the US has around 3,000 megawatts of geothermal capacity but almost all of that was in place by 1990 after a boom in the 1980s. In contrast, the Philippines currently has around 2,000 megawatts but its capacity is growing rapidly.

Researchers expect to be able to use geothermal plants both to provide low-grade heat and to generate electricity on a large scale. Geothermal has so much potential that the German government found in a 2003 study[1] (the "TAB" study) that Germany could cover half its electricity demand from geothermal plants. And that is not including heating, whose potential is two and a half times greater.

Keep in mind that Germany's potential is not nearly as great as that of more volcanic areas such as Iceland, the Philippines, and large areas in the United States. Germany is by no means the world's leader in geothermal power, nor will it ever be. The potential of other countries is far too great. Indeed, the US is currently the world's leader, just ahead of the Philippines, though geothermal provides less than 1 percent of the electricity in the US. Furthermore, geothermal is only the second largest source of renewable power (behind biomass) in the US and may soon fall into third position if wind energy continues to grow at current rates. But while geothermal power is concentrated in the western US, Germany's success despite its comparatively modest potential demonstrates that geothermal can be used everywhere, even outside areas with great natural geysers and volcanic activity.

The TAB study for the German Bundestag in 2003 found that the technical potential of geothermal power in Germany amounts to 300,000 terawatt-hours per year. Since Germany currently consumes 550 terawatt-hours of power per year, geothermal plants theoretically could provide enough power for 550 years of German consumption. However, Germany does not have to rely on geothermal power because it also has several centuries' worth of coal in addition to wind energy, solar energy, and biomass.

But, like biomass, geothermal can do something that wind and solar cannot: be ramped up and down to accommodate fluctuations in demand. And since biomass will be devoted largely to replacing oil for motive power, geothermal electricity plants will play a crucial role in replacing coal and nuclear power plants. Indeed, while critics of renewables point out that coal and nuclear power plants produce power most of the time — an estimated 60 to 70 percent availability — geothermal power plants are even more reliable at 95 percent.[2]

As we will see, the sustainability of geothermal power is not yet completely clear. No one is certain how long it will take for underground rock several kilometers deep to recover the heat taken out of it by geothermal plants. The German Bundestag report estimates that it will take "several centuries or even longer" and therefore lists the sustainable potential of geothermal electricity generation at 300 terawatt-hours per year, just over half of the 550 terawatt-hours of power the country consumes. In other words, geothermal power plants could cover Germany's entire base load.

How Does It Work?

Geothermal power is generated when heat is taken out of the Earth's crust to drive conventional turbines that generate electricity. This heat can also be used directly for heating and cooling. The temperature some 50 to 100 feet below ground tends to be constant year-round — generally around 60 degrees Fahrenheit, though temperatures obviously vary according to geographical and weather conditions. Simply running outdoor air through an underground system of tubes would allow us to replace our air-conditioning units by blowing this 60-degree air into our houses, a technology we will come back to below.

If we drill a bit deeper, the temperature below ground begins to increase, generally by about three degrees Celsius every 100 meters. We thus would have to drill only about three kilometers deep to pass the boiling point of 100 degrees Celsius almost anywhere. In the oil industry, such depths are no longer considered a challenge. In practice, boreholes have been drilled successfully some seven kilometers deep, and a borehole nine kilometers deep was drilled at a test site in Germany in the beginning of the 1990s.

In some areas, especially those with volcanic activity, the temperatures are far greater and the boiling point is reached much more quickly. The stretch of the Rhine between Basel, Switzerland, and the area southwest of Frankfurt, Germany, is one such region. In a project launched in 1987, a borehole was begun on the French side of the Rhine in Soultz-sous-Forêts. A few years ago, the borehole crossed the threshold of 200 degrees Celsius just short of four kilometers below ground. It is now more than five kilometers deep.

Hot Dry Rock

Such sites apply a method called Hot Dry Rock (HDR). A fluid (such as water) is injected into one borehole. It passes through natural fissures underground to a second borehole, where it is pumped back up to the surface. On its way through the hot, dry rock it absorbs the underground heat, which it carries back to the surface. Basically, underground rock is used as a heat exchanger.

In a project in Germany's Bad Urach near Stuttgart, a test borehole was drilled some four kilometers into the ground after initial tests at the end of 2002 had found that the underground rock could be used as a heat exchanger. The main risk in such projects is that the water will not make the trip from the injection borehole to the second borehole.

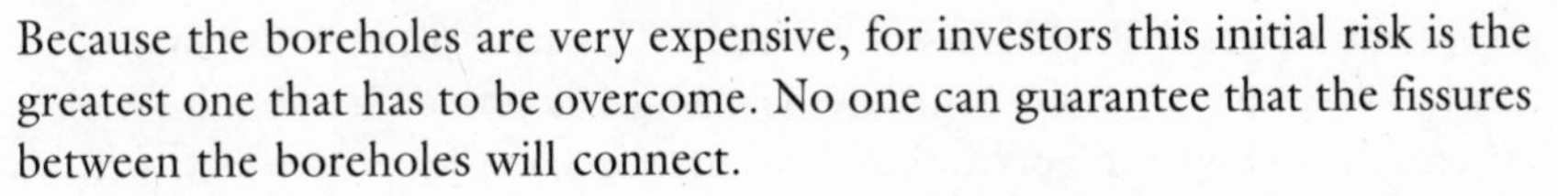

Because the boreholes are very expensive, for investors this initial risk is the greatest one that has to be overcome. No one can guarantee that the fissures between the boreholes will connect.

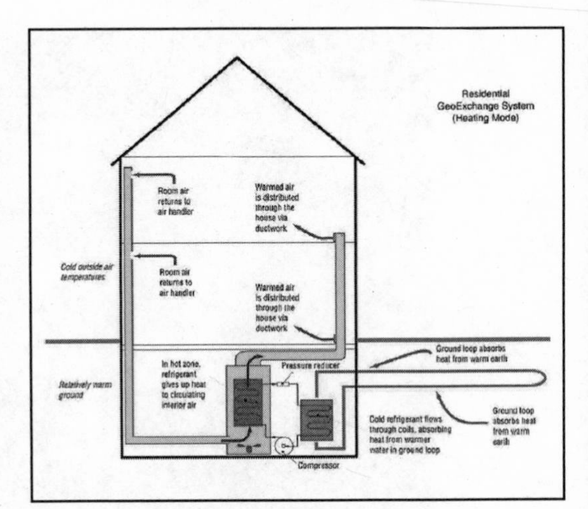

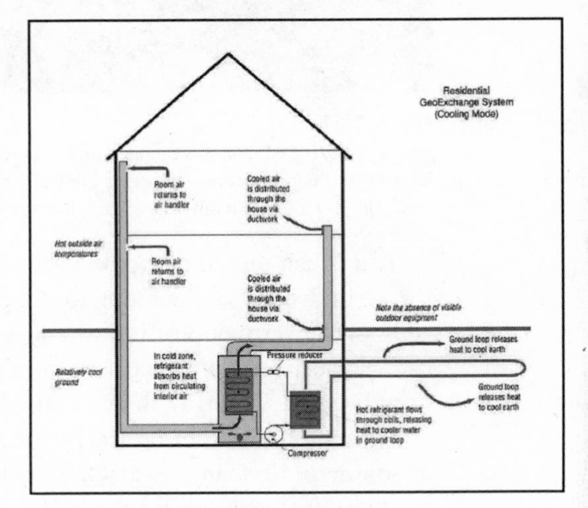

Figure 10.1a&b:
Geothermal systems. In geothermal heating and cooling systems, the difference between the outdoor air temperature and the relatively constant temperature a few yards below ground is utilized to heat or cool air that is blown into the house. Most of the energy needed here is consumed by the ventilation system, which means that such systems are much more energy efficient than conventional heaters or air conditioners. The ventilation system is essentially reversed to switch from heating to cooling. (Source: Geothermal Heat Pump Consortium)

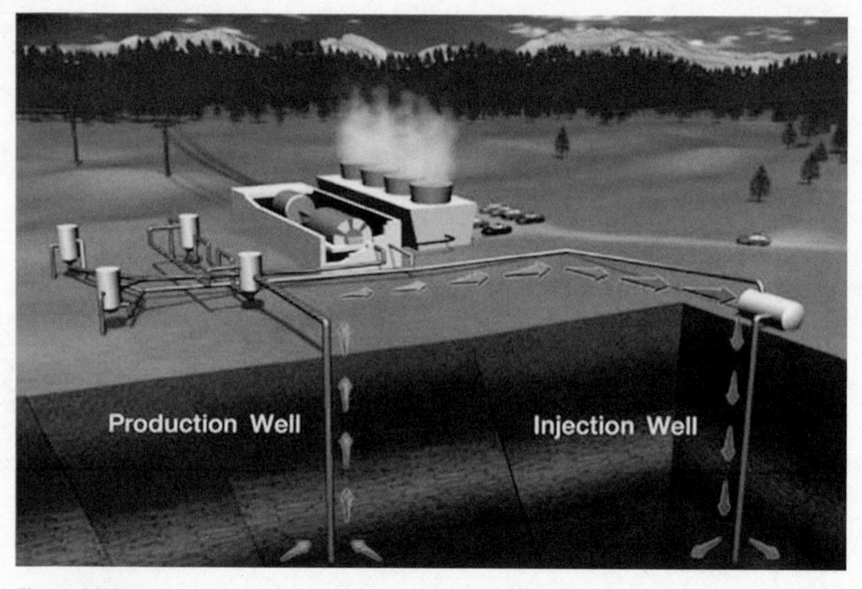

Figure 10.2:
Hot dry rock (HDR). Water is pumped into underground fissures through the injection well, where it passes to the production well and is pumped back to the surface, transporting the underground heat with it. (Source: Geothermal Heat Pump Consortium)

Bad Urach turned out to be a bad investment. In 2004, it became clear that the second borehole was not working as well as hoped and would require additional investments of at least 2 to 4 million euros — after 6.5 million euros had already been invested, most of it from the German Environmental Ministry. But Germany is just getting started as a producer of geothermal power and other recent projects there are running much better. Two projects in southern Germany — at Offenbach and Unterhaching, near Munich — produced temperatures far above 100 degrees Celsius in boreholes only about three kilometers deep.

It is estimated that HDR makes up some 95 percent of the geothermal potential in Germany. Aquifers make up the rest. Here, hot water is taken from below ground and used for heating or generating electricity. At a new plant in the town of Neustadt-Glewe in northeast Germany, an organic medium (perfluoropentane) whose boiling point is just below 30 degrees Celsius is used to generate electricity at temperatures below 100 degrees Celsius, thus allowing the use of underground heat to generate electricity in conventional steam turbines at much lower temperatures. This process is called the Organic Rankine Cycle (ORC). This new plant is the first of its kind in Germany, but the principle behind it has proved its usefulness elsewhere, for example in applications that use a second loop of a fluid such as

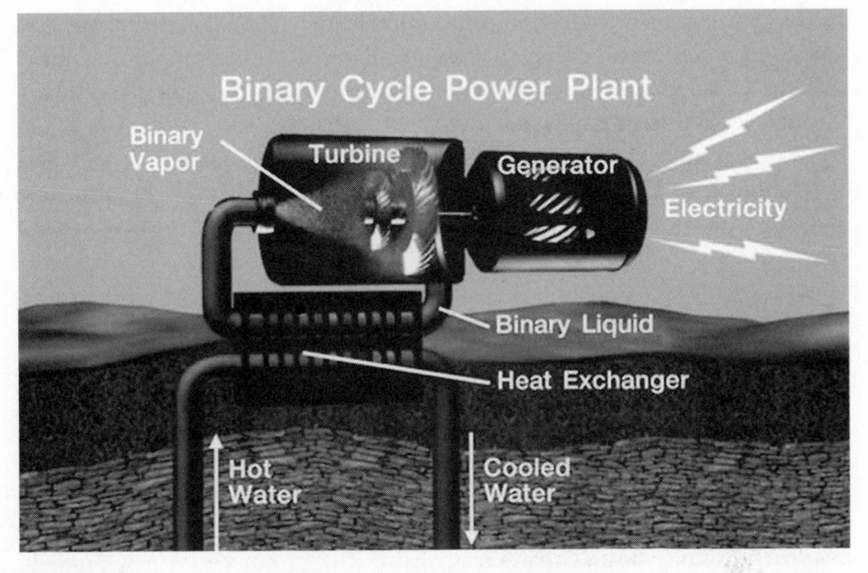

Figure 10.3:
Binary-cycle power plant. The binary cycle allows temperatures below 100°C to be used to generate electricity. Water is injected into the ground as with HDR, but if the temperatures are not above boiling at the outlet of the production well, the heat can be passed through a heat exchanger to a process fluid that does boil at the given temperature. Only water is circukated through the ground; the process fluid remains in a closed circuit. (Source: Geothermal Heat Pump Consortium)

ammonia (which has a low boiling point) to drive a steam turbine using waste heat from conventional plants.

Underground aquifers are also very useful for storing heat during the summer for later use in heaters during the winter. For a few years now, this approach has been put into practice in the residential area called Helios in the northern German town of Rostock. This process can be reversed for cooling purposes, as one project in Canada shows. The city of Toronto is taking cold water from Lake Ontario to cool down the air that is blown into downtown office buildings in the summer.

In the US, a similar heat pump system has been in use for many years in the Galt House East Hotel and the Waterfront Office Buildings in Louisville, Kentucky. This system cost only about half as much as a conventional air-conditioning system at the outset and saves another $25,000 a year in heating and cooling costs. In addition, the use of this geothermal system has meant that numerous heating and air-conditioning units are no longer needed and some 23,000 square feet of office space has been freed up. This system has been running smoothly for more than 20 years. The first major application for space heating in a district piped-heat network was in Boise, Idaho, where the drill can still be seen one block from the city's capitol building.

The small country of Iceland, with approximately 300,000 inhabitants, uses the most geothermal energy per capita, especially for heating. Icelanders already get some 86 percent of their heating energy from geothermal sources. Indeed, Iceland has so much geothermal potential that the country has begun growing tomatoes in greenhouses heated with geothermal energy. Because heat is so difficult to transport, Iceland is very interested in the development of a hydrogen infrastructure to allow it to export geothermal energy to continental Europe.

Where's the Hitch?

How environmentally friendly is geothermal power? No one really knows. In the beginning, researchers were not sure whether playing around with fissures in underground rock might cause earthquakes, especially since geothermal plants are generally put near volcanic areas. Researchers now agree that we do not need to be worried about this. As Dagmar Oertel, a co-author of the German Bundestag study mentioned above, explained to me,[3] the HDR approach exploits fissures that already exist; it does not create new ones. In addition, there is no danger of the ground sinking at the surface, which does happen above some oilfields. With geothermal power, nothing is taken out of the Earth; a heat conveyor medium merely circulates through it.

Nonetheless, some of the first geothermal plants near geysers in the US and New Zealand proved to be unsustainable. The chloride water was taken out of the ground and used in power plants but not reinjected into the ground. As a result, the sources went dry. In California, treated wastewater is now being injected into geysers in the largest geothermal field in the world (750 megawatts) to keep the field from drying out. The Geysers plant is not only the world's largest but also the oldest in the US, having been constructed in 1962.

This circulation is also important because the water from underground aquifers has an extremely high salt content (often as much as 20 times more than ocean water), which means that it cannot simply be pumped into the nearest stream. The same holds true for minerals and other substances that may be allowed to circulate through the power plant but should not be released into the environment. If any changes to the composition of the rocks or aquifers several kilometers below ground do occur, the biosphere at the surface probably will not be affected at all.

Thus the main problem that geothermal plants have to overcome is their price. In Germany, a kilowatt-hour from a geothermal plant (without consideration of the use of waste heat) costs less than 15 cents, but on the average that is easily twice as expensive as wind power and three times as expensive as electricity from coal plants. Only in a relatively small number of cases — in tectonically active areas such as the west coast of North America, in Iceland, or wherever aquifers are close to the surface and heat pumps are used for heating and cooling — is geothermal power competitive, with prices dropping even below 5 cents.

Ironically, geothermal power is used in low-grade applications in Germany's parliamentary building and on George W. Bush's ranch in Crawford, Texas. Bush has a heat pump to take the burden off his air conditioner. Maybe we should follow the example these politicians are setting for us. After all, coal power is getting more expensive, even if we do not include in the prices we charge consumers the extremely great external costs for the environmental impact of coal power. In comparison, the external costs for geothermal power are practically negligible. We need not fear the end of coal. We can have electricity — and we can make it clean. We just have to be willing to pay a little more.

Chapter 11

Fuel Cells

If you want to drive a Hummer, go to Iraq.
—*Thomas Friedman,* Grist Magazine, *2005*

Hydrogen: Much Ado About Nothing?

When people hear the words "fuel cell," most of them probably think of a fancy car with "NECAR" (New Electric Car) written on it. This vehicle has been around for a number of years but has yet to go into serial production. In 2005, Mercedes planned to build 60 of them. At the end of 2002, Toyota and Honda both brought fuel cell cars to market — with a leasing rate of $6,000 to $9,000 ... a month.

And yet fuel cells are nothing new. They were invented in the first half of the 19th century, and in 1874 French science-fiction author Jules Verne wrote of hydrogen as the "coal of the future."[1] Today, skeptics such as Walter McManus of JD Power and Associates believe that "hydrogen is the fuel of the future and it will always be the fuel of the future."[2] Others, such as former Shell CEO Peter Schwarz, believe that "by 2050, the world is running on hydrogen, or close enough to call it the Hydrogen Age."[3]

In 2003, Jeremy Rifkin saw it the same way in his book *The Hydrogen Economy.*[4] From fuel cells in cars to mass storage facilities for "excess" energy, hydrogen is said to have a number of benefits, but mostly it is touted as a substitute for oil without any environmental impact. Fuel cells that run on hydrogen are nearly silent and have no local emissions, making them excellent for applications such as forklifts that operate indoors. It sounds like a fairy tale: you pour in hydrogen and you get out heat, electricity, and pure water. Fuel cells are the key to this future hydrogen economy, which many claim will save civilization from the oil crisis.

But if you follow the news closely, you probably become more skeptical with each passing year. For many years, the automotive industry has been explaining that we will be able to buy fuel cell cars in ten years. Today, such cars can easily cost you $100,000 and you will pay $30 to fill up the tank — with which you can drive about 100 miles before you have to refill.

The applications are out there and they work. Will mass production reduce costs? Possibly, but — as we have seen with photovoltaics — feedstock can become scarce as production increases. Fuel cells powered by hydrogen require platinum, and prices for platinum have been skyrocketing in the past few years. But even if we are able to make inexpensive fuel cells, where are we going to get the hydrogen? After all, even fuel cells require fuel.

One crucial issue in the debate is whether the "new" electric cars using fuel cells are better than the "old" electric cars that run on batteries. Solar hydrogen not only would be a clean source of energy but also would solve the problem of what to do with electricity produced by solar plants and wind turbines when demand is low. In this scenario, "excess" electricity would be used to split water into hydrogen and oxygen in a process called electrolysis. But if hydrogen is to be used to store excess electricity, would it not make more sense to simply use batteries?

It turns out that the losses in electrolysis are great. In study after study, it has become clear that only around a quarter of the energy originally contained in electricity can actually be used in a fuel cell car, as Stephen and James Eaves argue.[5] In another widely read study,[6] researchers Roel Hammerschlag and Patrick Mazza agree: losses during electrolysis amount to around 10 to15 percent; additional losses occur if the hydrogen is compressed; and, finally, converting the hydrogen into electricity in the fuel cell takes 30 to 40 percent of the remaining energy, leaving only 45 to 55 percent of the original energy. Even without compression, the losses probably exceed 60 percent. (Keep in mind that the fuel cell itself is not the engine but only the source of power.) In comparison, the losses would amount to only around 10 percent if electricity were stored in lithium batteries. It thus makes more sense to store electricity in batteries than as hydrogen.

Taking this line of thinking a step further, researchers at the University of Warwick[7] have calculated that the UK would need 100,000 wind turbines or 100 nuclear power plants if all cars and trucks in the British Isles were to be powered with hydrogen from electrolysis. If wind turbines were used, the entire coastline would have to be surrounded by a 10-kilometer belt of offshore turbines, or an area the size of Wales would have to be covered with a wind farm.

No wonder studies have been showing that it makes more sense to invest in hybrid cars than fuel cell cars. A study conducted at MIT, for instance, found more potential in the development of diesel hybrid motors, which the researchers felt would be twice as efficient as fuel cell cars by 2020. Diesel engines can also run on biofuels.[8]

Such statistics, however, have not stopped some companies from promoting the idea of solar hydrogen. For instance, a VW technology center in Germany has a 50-square-meter PV system and a gas station that converts water into hydrogen via electrolysis. The system produces some 25 cubic meters of hydrogen per day, enough to take a car about 125 miles, so only one hydrogen car per day can fill up there. If we keep in mind that the 75 percent energy loss makes this hydrogen from solar power four times as expensive as the solar power already was, we can only conclude that VW wants to prove that solar hydrogen via electrolysis is an absurdity. Nonetheless, the company claims that the concept behind this pilot project will be ready for the market by 2015.[9]

Cost of 1kg of hydrogen	
From natural gas (reformation)	$2.00-2.50
From coal (town gas)	$2.00-2.50
From nuclear power (electrolysis)	$2.33
From biomass (pyrolysis, reformation)	$5-7
From a power socket (electrolysis)	$6-7
From wind power (electrolysis)	$7-11
From photovoltaics (electrolysis)	$10-30

Figure 11.1:
Cost of 1 kilogram of hydrogen. Hydrogen is far from being competitive with gasoline at the moment. Hydrogen is closest to being competitive when it is made from fossil fuels. But why convert fossil fuels into hydrogen when you can burn them directly? Most hydrogen is currently made from natural gas, so why not just fill up your car with natural gas? Note also that solar hydrogen is the most expensive of all. (Based on Timothy E. Lipman, *What Will Power the Hydrogen Economy?*, University of California, 2004)

This is not to say that fuel cell cars would not have some advantages over battery-powered cars. Above all, drivers would be able to fill up with hydrogen in just a few minutes, whereas it takes hours to recharge batteries. In addition, in mobile applications factors such as weight, safety, and range play an important role, and fuel cells often perform better here. But in terms of energy balance, battery-powered cars and hybrids are preferable to fuel cell cars. Think of it this way: as British journalist George Monbiot put it, "To fill all the cars in the US [with hydrogen via electrolysis] would require four times the current capacity of the national grid."[10]

In addition, the alleged problem of how to store electricity is not as great as the hype would suggest. Almost everywhere, it is very easy — though somewhat uneconomical — to simply ramp down power plants (gas turbines, coal plants, nuclear plants, and biomass plants) so that no excess electricity is produced. The problem of excess electricity in large grids is said to stem from the growth of intermittent renewables but in fact only wind and solar power are intermittent. Biomass and geothermal are not. (For tidal and wave power, see Chapter 12.) Indeed, the potential for solar power to be integrated into a grid is much greater than the potential for wind because, as discussed in Chapter 8, solar power production peaks around noon, when demand is generally the highest. In the end, it seems that only wind will occasionally generate large amounts of power when demand is very low (such as in the middle of the night), but as long as we do not get more than around 20 percent of our electricity from wind turbines we probably will not have much if any excess electricity.

At present, the only country that even comes close to having a fifth of its power from wind turbines is Denmark, though in northern Germany the share of wind in parts of the grid can be as much as a quarter. The experience of Denmark and northern Germany seems to suggest that 20 to 25 percent wind power is about all that a conventional grid can take. If the share of wind power rises above this level for a particular section of the grid, the grid will have to be modified. A study conducted by the German Energy Agency (dena) in 2004 on the integration of additional wind turbines in northern Germany, especially from planned offshore wind farms, found that more power lines would be needed to transport this extra electricity out of the north to the south, where less wind power is generated. This expansion of the grid would lead to considerably higher costs, but the study did not find excess electricity storage to be a problem.[11]

Black Hydrogen

Solar hydrogen would be a truly clean way of storing electricity. At present, however, many people probably believe that hydrogen is a clean *source* of energy. In reality, it is no more a source of energy than batteries are. Fortunately, many energy experts are starting to point out the difference between a source of energy and a storage medium. Coal, oil, and the sun are all sources of energy, while hydrogen has to be created using energy. When we convert natural gas (a source of energy) into hydrogen (an energy *carrier*), the hydrogen then constitutes secondary energy. When we convert natural gas into electricity (also an energy carrier) and then convert this electricity into hydrogen via electrolysis, the hydrogen constitutes tertiary energy. One simple analogy illustrates this difference clearly: it is still possible to dig a hole in the ground and find oil, but there is no hydrogen to be found anywhere on Earth.

Because most hydrogen produced today is made from natural gas and a large part of natural gas is made from coal, the future of the hydrogen economy is based on fossil energy. But if we keep in mind that not all fuel cells run on hydrogen we realize that fuel cells will not necessarily increase the consumption of fossil energy. Rather, fuel cells can also run on biogas and waste gases, such as the methane that traditionally has been allowed to escape unused from coal mines in order to prevent explosions.

We must separate the facts from the hype. Hydrogen-powered fuel cells do not have local emissions but, if they are made from fossil fuels, the emissions occur at the power plant or at the conversion facility. We therefore have to move beyond visions of people driving their own power plant to work and plugging in to the socket. US energy expert Amory Lovins of the Rocky Mountain Institute said, "The hypercar fleet will eventually have five to six times the generating capacity of the national grid."[12] Indeed, our current fleet of cars and trucks probably also has many times the generating capacity of the national grid, but no one has suggested that we get rid of our power plants and run the whole country on these potential electricity generators sitting around unused in our parking lots most of the time. So why are we suggesting this for fuel cell cars, especially since we do not know where we are going to get all that hydrogen?

Fuel cells do not, however, have to remain pie in the sky. We just have to make a distinction between the hyped-up hydrogen economy and the somewhat less sexy world of other types of fuel cells.

A Fuel Cell by Any Other Name ...

In 1839, William Grove developed the first fuel cell in Britain. The technology worked perfectly even back then, but when Siemens developed the dynamo in 1867 few industrialists saw any need for the fuel cell. Nowadays, as resources become scarcer, this old technology is making a comeback. Of course, the technology has changed considerably since Grove's time, though the basic principle remains the same. In the meantime, a number of different kinds of fuel cells have been developed and it is not clear which one will win out in the end. Perhaps different kinds of fuel cells will be used for different applications.

At the beginning of 2004, the US National Academy of Sciences published a study that caused quite an uproar. *The Hydrogen Economy: Opportunities, Costs, Barriers, and R&D Needs*[13] claimed that we were going to have to wait another 40 years for the hydrogen economy. Bad news for fuel cells? Not at all. Just bad news for hydrogen fuel cells — the kind people talk about most.

Low-Temperature Fuel Cells

PEM ("proton exchange membrane" or "polymer electrolyte membrane") fuel cells are the type most people think of. PEM fuel cells run on hydrogen and are used in cars. All fuel cells work more or less the same as a battery. In a PEM fuel cell, hydrogen is fed to the anode, with normal ambient air being fed to the cathode (the oxygen is used but it does not have to be pure for this fuel cell). The electrodes are connected to each other and current flows through the circuit. When it comes into contact with the catalyst — normally platinum in a PEM fuel cell — the hydrogen releases its electrons, which flow from the anode to the cathode, creating electric current. The protons from the hydrogen pass through the membrane to the cathode side, where they recombine with oxygen to create pure water.

Such fuel cells can be stacked together in any number to increase voltage and power output. In other words, these fuel cells are practically infinitely scalable. PEM fuel cells are used not only in large buses and cars but also recently in laptops. As PEM fuel cells have an operating temperature below 100 degrees Celsius, they can be used safely in portable devices. It helps that these fuel cells work efficiently under partial loads. They also are lighter than batteries with the same power output, making them interesting not only for mobile applications but also for large stationary applications such as providing emergency power to telecommunications base stations. Until recently, such base stations required quite large battery systems, which had to be supported by robust foundations.

But PEM fuel cells have one major drawback: the hydrogen has to be pure. Today, most hydrogen is reformed from natural gas. In the process, CO is produced — poison for the catalyst. At present, PEM fuel cells generally need a reformer to produce clean hydrogen from methanol or ethanol. PEM fuel cells require a hydrogen infrastructure that does not yet exist.

In recent years, researchers have therefore begun working on a type of fuel cell that uses methanol or ethanol instead of hydrogen: the direct alcohol fuel cell (DAFC). Here, methanol or ethanol is mixed with water on the anode side. Such fuel cells are interesting for very small applications such as laptops because the efficiency of DAFCs is lower than that of PEM fuel cells, which makes a big difference as the size of the system increases. However, at the moment it is a lot easier to refill with a small bottle of methanol than with hydrogen.

DAFCs have one major disadvantage: they emit carbon dioxide and are thus not as clean as PEM fuel cells. Both of these types of fuel cells belong in the category of "low-temperature fuel cells," which also includes alkaline fuel cells (AFCs) and phosphoric acid fuel cells (PAFCs). AFCs have been used for many years in space (such as on the Apollo missions) because they provide astronauts not only with power but also with water, a waste product of these fuel cells. AFCs are very well-developed, but the process gas has to be so pure that they are uneconomical for commercial applications. While PEM fuel cells make do with normal air, AFCs need pure oxygen.

PAFCs are the hottest fuel cells in this category, with operating temperatures around 200 degrees Celsius. They have one major advantage over all other types of fuel cells: they have been on sale for almost two decades. One such 200-kilowatt system provides electricity and heat for a police station in New York City. The largest PAFC unit — 11 megawatts — is in Japan. The major disadvantage of PAFCs is that the phosphoric acid solidifies at 40 degrees Celsius, making these fuel cells hard to start. In addition, they are only around 40 percent efficient. Researchers do not expect any major improvements here.

High-Temperature Fuel Cells

While CO poisons the membrane of a PEM fuel cell, solid oxide fuel cells (SOFCs) have ceramic electrodes that are not sensitive to CO. Another basic difference is that in SOFCs the charge carriers in the electrolyte are oxygen ions instead of hydrogen electrons as in PEM fuel cells. SOFCs can thus be run directly with syngas or natural gas because the CO is useful here, not poisonous.

This system also has disadvantages. Unlike PEM fuel cells, SOFCs are best run under full load. They cannot be easily ramped up and down because they have very high operating temperatures of up to 1,000 degrees Celsius. These high temperatures require very heat-resistant hardware. Currently, researchers are working to lower the operating temperature even as they increase efficiency. In 2004, researchers at Germany's Jülich Research Center managed to run a 40-centimeter SOFC stack with an output of 13.3 kilowatts — enough power for a multifamily dwelling — at 760 degrees Celsius.[14]

Molten carbonate fuel cells (MCFCs) are another type of high-temperature fuel cell. As the electrolyte in MCFCs is molten carbonate, these fuel cells run at temperatures of at least 650 degrees Celsius. High-temperature fuel cells are excellent as cogeneration units. The relatively high electrical efficiency of about 55 percent then increases to at least 70 percent. MCFCs require CO in the process gas; they do not run on pure hydrogen because the charge carriers are carbon ions. At the moment, the major drawback of this type of fuel cell is that the molten carbonate degrades the materials used so that such units do not have long service lives.

MTU CFC Solutions of Germany, one of the leading developers of MCFCs, has developed a 250-kilowatt unit it calls the HotModule.[15] Some of these units already have tens of thousands of smooth operating hours behind them. Several dozen HotModules are currently in operation in hospitals, universities, and industrial plants in Germany. Commercial production is scheduled to begin in 2006. These units had been running successfully for more than 20,000 operating hours at the end of 2005. HotModules are especially interesting because they can run on waste gas, mine gas, and other similar gases that have not traditionally been used commercially.

Are Only PEMs Clean?

We begin to see now why the hydrogen economy will not necessarily be clean. While it is true that the only thing that comes out of the exhaust of a PEM fuel cell car is pure drinking water, the emissions at the power plant are all the greater — at least until we find a way of producing pure hydrogen inexpensively and in large amounts without fossil energy carriers.

A lot of work is being done in this field. In late 2004, a team of researchers from Australia reported that it had developed a way of creating hydrogen from sunlight.[16] If this concept can be commercialized, it might be possible for homeowners to produce hydrogen on their roofs alongside their photovoltaic systems that produce electricity. This could truly represent a breakthrough on the path to a hydrogen economy. Researchers also are looking for a way of getting microorganisms to create hydrogen. This "biohydrogen" would be just as green as solar hydrogen. For many, such as John Heywood, director of the Sloan Automotive Laboratory at the Massachusetts Institute of Technology, moving the production of hydrogen away from fossil energy carriers is a prerequisite for the hydrogen economy: "If the hydrogen does not come from renewable sources, then it is simply not worth doing, environmentally or economically."[17]

In contrast, high-temperature fuel cells do not seem to be quite as clean as at first glance. After all, they run on fossil fuels. However, these fuel cells have a much higher efficiency than conventional coal plants. In addition, they can tap currently underused sources of fossil energy such as mine gas from coal mines, which would otherwise simply be allowed to escape into the atmosphere. The US Environmental Protection Agency estimates that less than 10 percent of the potential for the use of waste gas to generate electricity is currently being exploited.[18] High-temperature fuel cells could change that.

Furthermore, high-temperature fuel cells one day might also completely replace conventional coal plants if the coal is gasified. This process would make coal power much cleaner. In the end, the important thing is to keep an eye on the entire energy chain, not just what comes out of the tailpipe. Fuel cells could provide more electricity and heat from fewer fossil resources, thus making our energy supply more sustainable. This should be the measure of their success.

Chapter 12

Ocean Energy

I explored the coast along to the west and found there several flumes like the Kyle of Tongue, ready-made by Nature, through which the tide rushed, twice a day carrying thousands of tons of sheer power both ways. But nobody was doing anything about it. When I asked the engineers why, they said they did not know how to capture more than a negligible percentage of water power. I told them they had better find out .... But they went on grubbing for power in coal mines; and now that the atomic bomb [has] woken them up they are dreaming of nuclear energies, frightfully dangerous and enormously expensive ....
— *George Bernard Shaw, in a letter to* The Times, *February 14, 1947*

Wave power and tidal power are two emerging sources of renewable energy. Of all the sources covered in this book, they are the least developed, although a 240-megawatt tidal power plant has been in operation in La Rance, France, since 1966. The potential is great. According to George Lemonis of the Center for Renewable Energy Sources in Greece,[1] wave energy could cover 10 percent of current global electricity demand, with tidal energy potentially surpassing current demand by a factor of 50. After all, since the oceans cover about three-quarters of the Earth's surface, most solar energy is stored there. In addition, wave power and tidal power are reliable and predictable — two crucial benefits in a system that includes intermittent renewables.

Tidal Power

Why did the development of tidal power not continue after the success at La Rance? First, by 1966 France already had committed to nuclear power. In addition, the facility at La Rance is hard to reproduce. Situated in the naturally narrow and long crevasse between the cities of St. Malo and Dinard, the dam did not have to be very wide, even though the water volume is significant. Also, the plant doubles as a long-desired bridge, connecting the two cities directly. The largest tidal power plant (50 megawatts) in North America is on the Annapolis River in Nova Scotia.

The La Rance plant is essentially a dam like the common storage dams or run-of-river plants that have gone up all over the world, starting at Niagara Falls in 1886 and continuing today in the Three Gorges Dam now being built in China. Indeed, the turbines used in La Rance are the same as those used in run-of-river dams in France. La Rance has the potential to operate as a two-way tidal facility: the area behind the dam would be allowed fill up during high tide and drain during low tide. However, since the plant is at the mouth of a river, the basin partly fills up with water at low tide, reducing efficiency, and the plant now runs only in the ebb mode. If there were a lagoon instead of a river behind the dam, the plant would be more efficient in the flood direction, but, as we now know, the overall efficiency of such plants — called tidal barrages — is not significantly increased when they operate in both directions. To keep costs down, it thus makes sense to design a tidal barrage for simple generation during ebb. If geographic conditions permit, one tidal barrage could be built to produce power during high tide and another in a nearby lagoon during low tide so that the two plants would complement each other, feeding power to the grid at different times. When one plant peaks, the other stops producing altogether.

Unfortunately, as their design so closely resembles run-of-river dams tidal barrages have a noticeable environmental impact. Ecosystems develop around the roughly 12-hour tidal cycle, and tidal barrages cause the basins behind the dam to be filled with either more water than normal (for an ebb plant) or less (for a flood plant). Along with costs, this drawback probably is the reason other such projects — such as the one planned for the Severn Estuary in the channel separating Bristol, England, from Cardiff, Wales — have been put on the back burner. One of the best sites in the world for tidal power, the Severn Estuary site alone would probably cover 6 to 8 percent of Britain's electricity demand, with an estimated generating capacity of 8,640 megawatts (compared to 17,000 megawatts at the Three Gorges Dam). In 2005 there was talk of reviving the project, which was mothballed in the late 1980s when oil prices were low and Britain was awash in gas and oil from the North Sea.

New technology may be used there, for a lot has happened in the past few years. Tidal power has moved from dam-based designs to more modern submerged-rotor turbines, hydraulic lift-based devices, and pipes that use tidal pressure to drive air through turbines on land. The latter technology is being considered for use in San Francisco Bay, another of the world's best sites for tidal power, with the potential to cover more than the city's entire electricity demand. Here, HydroVenturi of Britain has designed a system that does without underwater electronic and technical equipment entirely: the water pressure pushes captive air in a chamber through turbines onshore.[2]

In contrast, a new tidal system based on "submerged wind turbines" takes advantage of the great energy density of water — some 850 times greater than air — to generate power with much smaller blades that also turn much more slowly than those of wind turbines. One of the major companies behind this technology is Marine Current Turbines Ltd, also of Britain.[3] Its SeaGen system, which has a rotor on each side of a pile and a generating capacity of 1 megawatt (roughly equivalent to the output of a mid-size onshore wind turbine), is based on the 300-kilowatt version called SeaFlow, which made do with only one rotor. Commercial production is expected in 2007 or 2008. These turbines could also be used offshore under decommissioned oil platforms, of which there will soon be many. (Indeed, there are also plans to have wind turbines atop such platforms where conditions are favorable, such as off the coast of Louisiana.)

Figure 12.1:
The SeaFlow in action with its rotor raised above water. The new double-rotor SeaGen will have one rotor on each side of the pile. (Source: Marine Current Turbines Ltd)

The Stingray concept is completely different.[4] Here the tidal current pushes a wing up and down, pressurizing oil to drive a hydraulic motor that powers a generator. After several test modules proved successful in pilot projects off the

Shetland Islands, the British firm Engineering Business is now testing 500-kilowatt units. One obvious benefit here is that the complete lack of fast movement will eliminate any danger to sea creatures.

Figure 12.2:
Stingray (Source: Engineering Business Ltd)

It is, of course, no wonder that Britain is so active in the field of tidal energy. Britain has the most tidal power potential in Europe, ahead of France and Ireland. But given the mere size of the coast of North America, the US and Canada cannot afford to miss out on this development. After all, Germany has very little coast and no great tides but that has not stopped Germany from investing in the SeaFlow project.

Wave Power

Three types of units have been developed to generate power from waves. One uses overtopping water, the second uses an oscillating water column, and the third is a pitching device.

The overtopping system is called the Wave Dragon.[5] A joint project of six EU member states, the Wave Dragon was the first wave energy system connected to a grid. The unit is anchored to the sea floor but floats at the surface. Waves are propelled up a ramp to the reservoir in the middle, where the water falls through turbines to generate power. Two 30-meter-long Wave Dragons now produce power off the shore of Denmark around 80 percent of the time at a cost of 7 to 10 cents per kilowatt-hour. It can be used between the 30th and 60th latitudes, where waves are sufficient. A 150-meter-long unit with an output of 7 megawatts is scheduled to go into operation in 2006 off the coast of Wales.

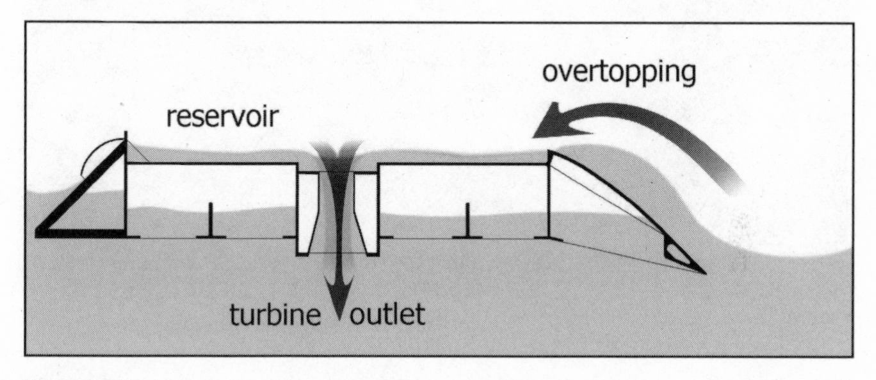

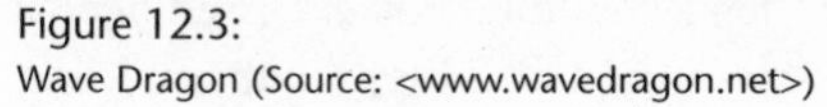
Figure 12.3:
Wave Dragon (Source: <www.wavedragon.net>)

Wavegen of Scotland is behind the latest oscillating-column system (OCS).[6] Here, an air chamber is created in a facility directly on the shore and the pressure in the chamber increases and decreases as waves impact. The resulting push and pull on the air inside the chamber drives a Wells turbine on land that rotates whether the air is passing in or out.

Finally, Ocean Power Delivery Ltd of Scotland has developed the pitching device called Pelamis.[7] The first commercial wave farm constructed off the coast of Portugal in 2005 combines three 750-kilowatt units. Each of the 150-meter-long units consists of five cylinders hinged together. Their motion against each other is used to pump fluid through hydraulic motors to drive generators, much as is done in the Stingray. To develop its excellent resources, Portugal currently is paying 23.5 cents per kilowatt-hour for electricity from wave energy.

The array of possible solutions is great, and it is still quite unclear which will be a commercial success first. For example, Wavegen's OCS plant fell somewhat short of the target output, though the engineers state that the shortfall was due to insufficient knowledge of the site rather than any inherent problem in the system.

Figure 12.4:
Pelamis (Source: Ocean Power Delivery Ltd)

Indeed, the winner may even be another technology not discussed here at all — perhaps an ocean thermal energy conversion (OTEC), or thermocline, system. Here the temperature difference between surface ocean layers and deeper water is utilized to generate power, for instance in a Rankine or Kalina Cycle. Temperature differences in ocean layers are greatest at the equator, making this solution a good complement to wave power, as waves are weak at the equator. The first such unit, developed by the Frenchman who invented neon, went into operation in 1930 in Cuba. It did not pay for itself — more power was used to pump water to get the process started than the plant generated — and Georges Claude soon went bankrupt. More than 75 years later, not enough research has been done to produce a plant that would be economical, even though its principle has been demonstrated to work.

Demand Management

New technologies are already being used to control output from distributed generation at several sites to respond to market conditions, creating a kind of "virtual utility." The operation of a network with a large number of virtual utilities will require much greater real-time information flow than is now the case.

—*International Energy Agency,* Distributed Generation in Liberalised Electricity Markets, *2002*

Virtual Power Plants

Because solar and wind power are intermittent, there is much talk about how to accommodate fluctuations in power supply. Usually the argument is that, even if we develop wind power to a great extent, conventional plants — coal, nuclear, and gas — will still be needed to meet peak demand when the wind is not blowing.

Opponents of intermittent renewables often argue along these lines, as exemplified by Erwin Teufel, governor of the German state of Baden-Württemberg:

> To ensure a power supply, we need what is called "reserve capacity." In practice, this basically means conventional power plants that can produce more electricity when the wind is not blowing. If Germany continues to expand its wind power capacity, we will basically still end up having to expand the generating capacity of our large power plants, including those that cover the baseload. Otherwise, we will not be able to prevent the grid from overloading or collapsing. In turn, these conventional power plants will then often have to run below full capacity, which means that they will not be as efficient as they could be. [1]

This solution addresses only the supply side. The drawbacks of the argument are obvious. When people question whether wind power actually reduces carbon dioxide emissions at all, this is the line of thinking they are following. It is not unfounded but it is also not creative.

The approach from the demand side goes by various names: demand management, intelligent grids, and virtual power plants. The distinctions among these still poorly defined terms are blurry, and some are also used to refer only to the supply side. Indeed, while the German term *Ivirtuelles Kraftwerk* (virtual power plant) is commonly used, the variety of terms in English may be the result of the trademarking of the term "virtual power plant" by Encorp, which also understands the term to refer to supply-side management. The company's website states:

> The Virtual Power Plant™ links together seldom-used standby and emergency backup generators at hospitals, universities, manufacturers, office towers and other facilities. This network of "mini" power plants allows utilities and high-energy users to draw additional power from these on-site sources as needed.[2]

In this chapter, I will be using the term "virtual power plant" to mean control of both supply and demand, as I did in the German version of this book. In light of the recent power outages in North America and Europe, we need such solutions today. There is no need to wait until intermittent renewables make up a large share of our power supply. Power grids are already in trouble and distributed generation with demand-side management is the answer.

How does distributed generation (DG) work? To begin with, utilities are not always in favor of DG, which consists of small power generators, usually of only a few megawatts or even smaller — solar panels on roofs, wind turbines, and also diesel and biomass generators, cogeneration units, and fuel cells — as opposed to large, central plants with hundreds of megawatts. Power from such small units is generally generated close to the consumer, quite often in the same building. This power does not have to go very far to the grid to reach the consumer, if it enters the grid at all. Distributed generation thus takes a load off the grid.

While some utilities are trying to argue that they should not have to pay so much for power fed to the grid from small, distributed generators because they allegedly cause uncontrollable power surges, studies have shown that DG can actually improve power quality considerably. One such study recently was conducted by New Power Technologies of California and was funded in part by the California Energy Commission.[3]

A 2004 report, "Economic Growth and the Central Generation Paradigm," also found that costs can be saved if we switch from central to distributed generation:

> The world can save $5.0 trillion or 46% in capital costs by building all new generation near users. By avoiding line losses, recycling industrial process heat to produce power and recycling waste heat from fuel fired power generation, the DG approach avoids use of 122 billion barrels of oil equivalent of fossil fuel, saves $2.8 trillion in fuel costs and cuts carbon dioxide emissions associated with incremental power generation by 50%.[4]

Overall efficiency would increase under DG. Thomas R. Casten and Brennan Downes point out that grid efficiency peaked in the US in 1910, when much power was still DC from local generators. They add:

> No other industry wastes two-thirds of its raw material; no other industry has stagnant efficiency; no other industry gets less productivity per unit output in 2004 than it did in 1904.[5]

What could utilities have against using DG to improve quality and lower costs? Leaving aside the political issue of control, the market has traditionally failed to promote efficiency in the design of compensation for public and private monopoly utilities, which are often required to pass on savings to customers in return for guaranteed profits, Since costs too can be passed on, there is often no profit incentive for utilities to invest in saving power.

Even if utilities eventually welcome more distributed generation, there are still a few problems to solve. One is that the power a single generator feeds to the grid should come at a time when the grid needs more power. Utilities therefore must send signals to these generators to let them know when to increase output. Obviously, with solar panels and wind turbines this is not possible. But most other generators would be able to increase output if they received a signal to do so. That is the supply side. Can we also send such signals to consumers to manage demand?

Imagine that your refrigerator has a small display and is connected to the Internet. Dear reader, you think you know what I am about to say: the refrigerator reorders milk automatically. No, in my scenario you will be walking or biking to the grocery store to get your food yourself; Americans need more exercise. The refrigerator I am talking about tells you how much power from the grid costs at the moment. The price is a reflection of how much generating capacity is available in central power plants. This price, which could change by the minute, would drop if there is enough capacity on the grid and rise when fossil power is scarce.

You could then set your refrigerator to automatically switch off if a certain price is reached. If your refrigerator generally is set at, say, 45 degrees Fahrenheit and switches off when power rises above a certain price, you could have it turn back on regardless of the price once its temperature rises above 50 degrees. It could also cool down to 40 degrees if the price drops below a set level. Other appliances could run intermittently along the same lines, such as washing machines, which could be set to switch on when power reaches a certain price, and laptops, which could be set to switch to battery power when the price rises to a certain level.

In one such program in Austin, Texas, the local utility company Austin Energy offers consumers a financial incentive if they allow the utility to switch off their air-conditioning units during peak demand. This arrangement is very useful for the utility company, which does not have to build additional power plants in order to meet peak demand, while the sacrifice for consumers is negligible because the temperature in their homes rises only minimally for a very short period on rare occasions, generally only a few times a year.

The potential here is great. According to the Electric Power Research Institute, if California had shaved its peak consumption during the blackouts of 2001 by 2 percent, wholesale electricity expenditures would have dropped by $700 million.[6] The Brattle Group has calculated that a 10 percent reduction would have cut wholesale prices in half.[7]

What kind of signal do we need? Take your pick. There are all kinds of technologies already available for this purpose, but most of them are currently being designed or used for more "light-hearted" applications (such as reordering milk, as though that were a problem). For instance, the Ambient Orb is an egg-shaped device that receives signals and changes color depending on whether the signal is increasing or decreasing.[8] The Orb is programmed — just what the world needs — to change color to indicate whether the Dow Jones is rising or falling, but the manufacturer says it could just as easily show whether the temperature in New York is rising or falling if there were a signal for that. If utilities decided to send out a signal to distributed generators indicating that they should start switching on or off, this technology could easily be used.

The environmental website *Grist Magazine* reported in 2003 on a technology by Sage Systems called Autonomous Local Area Distributed Network (Aladn) that "allows customers to go online to monitor and adjust the energy consumption of home devices."[9] The company's website has apparently been offline since 2003 and little has been heard of this system since. The Aladn website is also offline. Most likely, the technology worked just fine but the company was unable to find the right market. At the Home Toys website (the name speaks volumes), an article from 2000 tells us what was envisioned:

> Aladn enables remote Internet access of devices in a home, safely and securely. With the proper password, a homeowner can turn on the heat in a snow-bound vacation home before driving there, or receive an e-mail at work when the kids arrive home from school.[10]

In other words, the company wanted to market its technology as a toy for grownups rather than as an energy-saving device.

All sorts of other Internet-based technologies could be used. For instance, the 77 members of the Open Service Gateway Initiative, founded in 1999, include companies such as Ericsson, IBM, Philips, Sony, and Whirlpool.[11] This initiative uses existing service gateway technologies to network intelligent devices. Recently, the HomePlug Powerline Alliance was created to promote a specification for the networking of appliances.[12] Another system, previously called iReady and now being marketed as nChip, embeds Internet technology in appliances.13 Such devices could then be switched on and off automatically depending on the price of a kilowatt-hour — if only there were a signal.

Actually, such signals are already being used in the energy sector but they are not connected to appliances. In a system called EFR (*Europäische Funk-Rundsteuerung*, or European Radio Control),[14] utility companies send commands via ISDN lines to a central station, which sorts them by urgency and passes them on to two transmitters, one outside Berlin and the other near Frankfurt. These two antennas cover all of Germany and parts of nearby countries. Cities like Dresden and Berlin already use these signals to switch their street lights on and off. The same technology is used in the southern town of Augsburg to turn the street lights on for the weekly meeting of skateboarders; a simple laptop can be used to switch the lights on and off. EFR is also used to switch heaters on and off in the southern German village of Sindelfingen to prevent peak loads in gas lines, for instance in the morning when everyone gets out of bed and turns up the heater. As the name says, this system is designed to be expanded across Europe.

After the power outages in the summer of 2003 in North America, a manager at Infineon spoke of the danger that arises after a power outage when appliances that remained plugged in and switched on suddenly receive power.[15] The newly juiced grid then receives a quick shock as all of these appliances immediately drain power from the lines. The results can be an immediate second power outage. In theory, the same thing can happen with demand management. If everyone sets their refrigerators to switch off when the cost of the kilowatt-hour reaches 15 cents, for example, there will be a sudden drop in demand that will reverberate through the grid, potentially raising the hertz frequency to levels that could damage appliances that are still connected. The price would then immediately drop, sending out a second signal for all devices to switch on again — possibly with similarly disastrous results. Obviously, if we are going to have demand management, we have to find a way of preventing such feedback. Cascading is one obvious solution. Central control of appliances could ensure that consumers are switched off not all at once but consecutively. And if people react manually to rising prices on a display updated by the minute there is little danger of too many people acting at once.

Demand management is included in the state of Pennsylvania's renewables portfolio, implemented in February 2005.[16] Electric City, a company specializing in energy efficiency technologies, has developed what it calls a Virtual Negawatt Power Plant.[17] (For a discussion of negawatts — saved energy — see Chapter 14.) The company already operates two demand response systems in the US: a 50-megawatt system in Chicago and a 27-megawatt system in Utah. Such systems demonstrate that demand management and efficiency go very well together. And while much is happening in North America in this field, Europe is not missing the boat either. Electricité de France has come up with a similar system it calls Intelligrid, which it has even exported to New York.[18]

The Electric Power Research Institute speaks of an "energy Internet" in which distributed generators work together, with the intelligence coming from the grid itself.[19] Similar projects go by the name SmartGrid,[20] and the US Department of Energy has founded the GridWise Alliance[21] to further develop the basic idea behind virtual power plants. In addition to hourly rates, there is the concept of interruption rates, such as when customers receive signals that the power supply is getting tight so they can manually switch off non-critical appliances such as electric heaters and air conditioners.

In the US, the Demand Response and Advanced Metering Coalition (DRAM)[22] praised the Smart Metering Section of the Bush administration's Energy Policy Act of 2005 for including a requirement that customers be allowed to have time-based rates, with meters displaying current rates. The Section also requires governmental monitoring of progress. In a program to be completed in 2006, the US Department of Energy is developing an EnergyNet,[23] which will allow consumers to be controlled via the Internet to shave peaks and switch on UPS generators when more power is needed.

Empowerment

In this scenario you not only are a power consumer but you also operate your own generator. You own have solar panels on your roof (a 4-kilowatt system of attractive solar tiles — just a bit larger than the 3.8-kilowatt system your now envious neighbor recently installed) and shares in a local wind farm and a fuel cell in your car. Your heater — if you still need such old-fashioned technology because you have not yet moved into a Zero Energy Home (ZEH) — consists of a small biomass-fired cogeneration unit in your basement that generates electricity and provides enough waste heat to cover all your hot water needs. You can not only switch off appliances to reduce demand but you can also generate more power and feed it to the grid when the grid needs it and prices rise. At least, so goes the theory.

In reality, the electricity industry does not like to hand over control of its natural monopoly to the consumer.[24] Indeed, returning to the Open Service Gateway Initiative we see that the idea that consumers could be empowered is not very clear even to industry researchers. As one researcher put it in 2001:

> If we take these visions to their natural conclusion, we could have a washing machine that users rent or lease directly from their utility company. The rates paid would then be based on when the washing machine runs, which would be determined via networking.

The researcher does not explain why such appliances should not simply be sold to consumers. However, my vision of consumer empowerment is shared by some in the industry. Take a look at what Siemens wrote in its corporate magazine in the spring of 2002:

> ... the telephone rings. On the display, a young man appears and introduces himself as a representative of a new energy dealer. Dost knows the company; they're the ones with the stop-and-go power light. "Mr. Dost, don't you, too, have the feeling you pay too much for your power?" says the sales rep, beginning his pitch. "We have a completely new concept: With our stop-and-go power light, you can always see when it's an especially good time to mow the lawn or turn on the washing machine. If the light is green, the power is at least 20% below the spot price on the energy exchange. All we have to do is install a small program in all of your Internet-connected household appliances, and without further ado you will be able to enjoy our benefits, Mr..."
>
> "Excuse me," Dost interrupts. "Our residential community has leased a 100-kW fuel cell with a microgas turbine; we produce electricity and heat and sell the surplus to the distributor or the power exchange, depending on which one pays the better price. But

maybe you have a piece of software that will improve our network management? Call back again tomorrow, if you would, just now isn't convenient." [25]

Chapter 14

Efficiency

> The best strategy would involve a mix of actions on energy efficiency, including conservation measures, renewable energy and switching to low CO_2 fuel and gases. Perhaps half of the potential growth in emissions could be saved by greater energy efficiency.
>
> —*Joachim Faber of the German banking group Allianz, 2005*

Negawatts: Saved Energy is Our Greatest Source

In calculations of future energy demand in which renewables play a large role, efficiency often is the major factor. In many cases, the technologies needed are already available. In other cases, no technology is required at all; we simply have to change our attitudes and behavior.

In all cases, the changes required often do not work because of human failure. Too many people feel that creature comforts are more important than saving energy and too few people understand that they can save energy without giving up these comforts. In addition, prices are often mistaken for costs.

Major advances in efficiency are possible overnight. For instance, it has been estimated that the French could shut down one and a half nuclear power plants simply by switching from incandescent bulbs to energy-saving lamps. Energy-saving lamps cost far more than incandescent bulbs but, since they save a lot of energy and stay in operation much longer, they pay for themselves several times over.

How many consumers understand that? Some oppose governmental enforcement of efficiency; they want to leave everything up to markets. Sterling Burnett of the National Center for Policy Analysis in Dallas summed up this philosophy when he stated, "I have problems with government forcing these choices through tax policy, subsidies or mandates"[1] — in other words, it is up to consumers, and government should not be dictating efficiency and conservation. But even on those rare occasions when consumers do know the total cost of ownership of appliances, the market often erects obstacles. Take the example of air conditioners. Lowering peak consumption by using efficient units would save

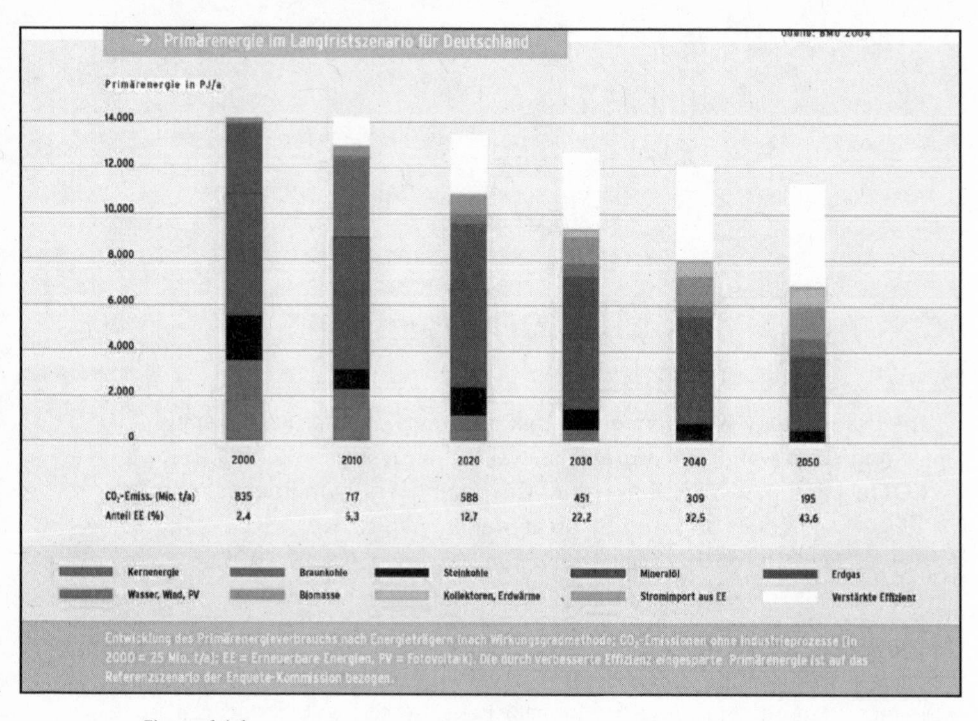

Figure 14.1:
A scenario for a 43.6 percent supply of energy from renewables in Germany by 2050 (up from 2.4 percent in 2000). Here, efficiency (the white section of the bar) meets around 40 percent of energy demand, the largest share. Note that this chart assumes energy demand will drop by about a fifth by 2050, from just over 14,000 petajoules per year to slightly more than 11,000 petajoules per year, even before efficiency reduces actual energy consumption to below 7,000 petajoules per year. In contrast, nearly every official US governmental forecast — not to mention the forecasts of energy companies and many international organizations — has assumed consumption in the US would increase. If it increased at 2 percent per year, energy consumption in the US would almost triple over 50 years. The German government plans to cut consumption in half by then. (Source: German Environmental Ministry)

utilities the enormous costs of adding generation capacity. Consumers would save money when utilities kept prices stable and the extra investment would pay for itself. The Bush administration attempted to reduce the Seasonal Energy Efficiency Ratio (SEER) from 13 to 12 because manufacturers could then sell cheaper units, but utilities would have been forced to expand capacity. Here the market did not support efficiency; a court had to overrule the attempted change.

In Freiburg, Germany, energy expert Dieter Seifried convinced the local utility company back in the early 1990s that it would be cheaper to hand out one free energy-saving lamp to every household than to construct another power plant to cover demand. The plan worked. It also had a pleasant side effect: it put this new technology into every home, allowing people to see what the differences were between this new type of light bulb and the old ones (sometimes a slight delay in illumination; little heat given off by the bulb, which can be touched while on; and a slightly different light color).

Now compare how energy-saving lamps were being marketed in the US in the 1990s. On a TV commercial, a man was shooting hoops in his driveway at night by artificial light. A voice off-screen explained, as the man missed one shot after the other, that if he kept practicing he probably would need to replace the light bulb someday. No mention was made of energy savings and total cost of ownership. Apparently the manufacturer thought that the only way to market the long-lasting light bulb to Americans was to point out that it was good for hard-to-reach sockets, where bulbs are a pain to change.

Of course, there are areas in which saving energy means changing one's lifestyle. Gasoline and diesel engines are well-advanced. The main increases in fuel efficiency here stem from making cars with smaller engines, lighter frames, and better aerodynamics, not better motors. Indeed, if Americans simply started driving the more efficient cars already on the road in Europe, the US could do without oil imports completely. But then Americans would be driving small cars, not giant off-road vehicles on the street.

But Europeans are also not driving with the most advanced technology — far from it. For a few years in Europe, Audi sold a car it called the A2, which got almost 80 miles per gallon. When I lectured in Austin, Texas, in early 2005, someone asked if the figure "80 mpg" on one of my slides was a typo. "No," I reassured her, "it's not a typo; it's a Lupo." Volkswagen had a version of its Lupo on the market during the same period as Audi's A2 with roughly the same gas mileage. Both these cars were taken off the market completely in mid-2005 even though gas prices were roughly twice as high as in North America. Europeans calculated that they would recoup their initial extra expenses only over the lifetime of the car. The Lupo and the A2 were not hot items.

Unfortunately, Germany's tax system inhibits purchases of such vehicles. Drivers are able to write off their car expenses in full after 1 percent of the initial list price of the car is deducted per month. Let us say, for instance, that you own a car with a list price of 10,000 euros and you spend 150 euros per month on fuel, insurance, and other expenses. You can write off 150 euros minus 100 euros (1 percent of the list price), leaving you with a write-off of 50 euros a month. If you purchase a slightly more expensive car — say, 12,500 euros — in order to lower your monthly expenses for fuel by 20 euros a month, the tax system works against you because the equation is now 130 euros in expenses minus 125 euros, leaving you with only 5 euros a month to write off.

There is also a skewed tax incentive in the US for the highly criticized off-road SUVs, which in many states are taxed not as private cars but at the lower rate for light trucks. When this highly questionable tax incentive made headlines in 2002, it seemed for awhile that the law would be changed, but in 2003 the incentive was widely maintained and even increased in some places. To make matters worse, when the then environmental secretary, Christie Whitman, appeared on the Diane Rehm show (a radio call-in program on NPR) in May 2003, [2] she did not seem to be aware of such tax loopholes for SUVs when callers demanded that she explain the environmental sense behind such legislation.

Lupos, A2s — have you heard of these cars? They were not sold in North America. Apparently the companies did not feel there was a market for them. Indeed, even some of the most common cars in Europe are not sold in North America. When you rent a car in the "economy" class in the US you may find yourself driving a Dodge Neon with 135 horsepower. My car in Europe is a Renault Twingo — a true economy-class vehicle with 55 to 75 horsepower, depending on the model. The Dodge Neon gets between 30 and 35 miles per gallon but a Twingo easily gets above 40. The Twingo has a top speed of only around 100 miles per hour, but that is more than the law allows almost everywhere in the US. Americans not only like to drive off-road vehicles on the road but also seem to want cars built to get them speeding tickets. Somehow all of this is a luxury that few want to do without. Although parents often tell their children they can't have something because they simply don't need it, adults do not seem to apply this rule to themselves.

Even in Europe, as we have seen, technology is far ahead of what people consider acceptable. Not only have the two most efficient cars been taken off the market but also the prototype of a car that would get closer to 250 miles per gallon — a two-seater by Volkswagen — has been mothballed indefinitely. Indeed, the 40 miles per gallon of the Renault Twingo is nothing compared to the over 70 miles per gallon one could get with the Twingo if Renault would implement the modifications Greenpeace made to the car they called SmILE (Small, Intelligent, Light, Efficient). But, alas, unlike the CFC-less refrigerator Greenpeace designed and industry later adopted as the standard model currently sold in Europe, Greenpeace's Twingo has not been included in Renault's product line.

Even simple changes in behavior can increase energy efficiency. Conservation experts advise us not to preheat our ovens (except when baking sensitive cakes) and to switch off electric burners, which remain hot for many minutes, before we finish cooking. We waste a lot of electricity by leaving all kinds of appliances and devices on standby. Germany's Environmental Ministry estimates that each German spends around 42 euros per year on electricity just to leave televisions, answering machines, and such on standby all the time. [3]

The Jevons Paradox

At least since the publication of *Factor Four: Doubling Wealth, Halving Resources*[4] in 1995, it has been clear that efficiency is no longer a technical challenge. More often than not, the challenge, if not merely a question of attitudes and awareness, is economic.

The economic challenge has two aspects. If utilities make money by selling power, how are we going to get them to promote efficiency and in addtion, efficiency lowers demand, which in theory lowers prices, and cheaper power would entice people to consume more. As the Jevons paradox puts it, [5] increasing efficiency increases consumption. Or, to put it differently, the higher prices are, the more efficiency pays for itself.

Let us come back to the program in Freiburg, Germany, where the local utility company handed out energy-efficient lamps to reduce power demand. Since this program was driven by the need to avoid construction of a new plant, the utility welcomed the proposal. But if consumers simply decided to start investing in efficient equipment, the utility would be facing a problem because certain fixed costs remain the same. Events on the water markets in eastern Germany since the fall of the Berlin Wall illustrate this effect clearly, though the cause is different. In the past 15 years, cities there have lost as much as a third of their populations, with Halle dropping from around 300,000 to 200,000 inhabitants as people migrated to the west to find work. Overnight, the municipal works were faced with a problem: fewer people were buying water but the fixed costs did not change. The result was an increase in utility prices in the very areas that were suffering from high unemployment and economic depression.

Dieter Seifried, the instigator of the energy-efficiency plan in Freiburg, studied such issues in his book *Das Einsparkraftwerk*[6] (co-authored by Peter Hennicke) with a case study of the municipal works in the city of Hanover. The authors found that the city could save some 40 megawatts if efficiency incentives were implemented, but that these measures to save energy would cause the utility company to enter the red. Their solution was quite simple: the utility should be allowed to cover its losses by raising prices.

This solution would kill two birds with one stone. On the one hand, the utility companies would have no real reason to fight the implementation of energy-saving programs, for they would have nothing to lose as this policy is revenue-neutral for them. On the other hand, higher prices ensure that investments will pay for themselves. At the same time, consumers would still be able to save money: although the price for a kilowatt-hour of electricity would increase, consumers would be able to consume so much less power that their costs would be lower in the end.

Of course, we can wait until the scarcity of resources reduces supply, which also increases prices and makes efficiency a good investment. Why not wait? Let us take the case of petroleum. The United States is currently importing around 60 percent of the oil it consumes. This money leaves the country. Some economists claim that this does not matter because the oil-exporting countries

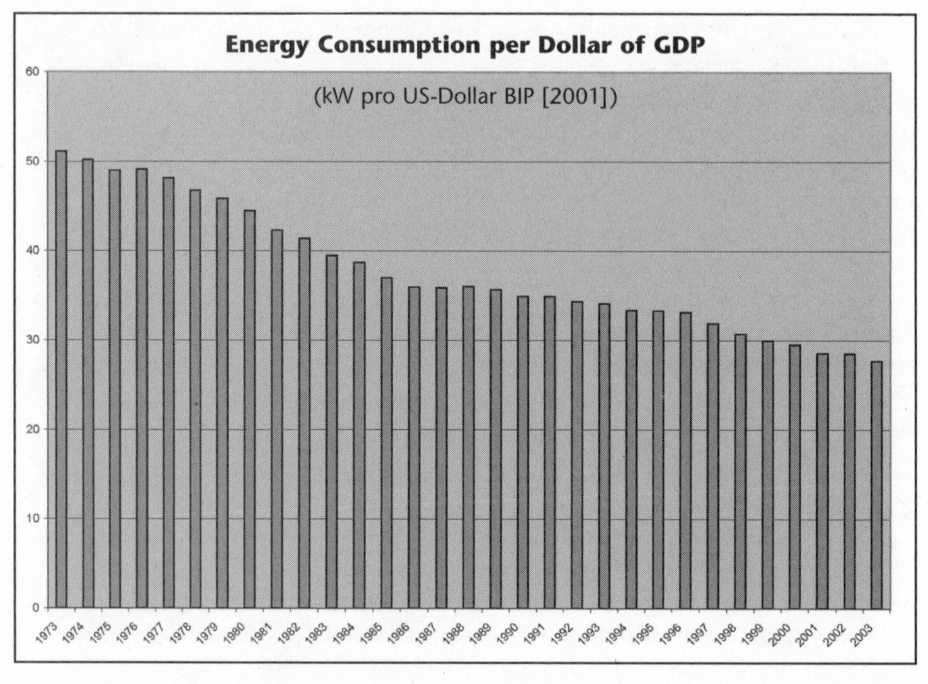

Figure 14.2:
Energy intensity (kilowatt-hours per dollar of GNP) in the US. Since the first oil crisis in 1973, energy consumption in the US has dropped relative to gross national product. In other words, more and more product is created with less and less energy. The figures above show that this "energy intensity" in the US dropped by almost 50 percent from 1973 to 2003 (adjusted for inflation, with the reference value being the dollar in 2000). Indeed, all the OECD countries have cut their energy intensity in half since 1973. This trend continues today. (Source: US Department of Energy)

will then turn around and buy US products with these dollars. Unfortunately, this argument is such an oversimplification of the role of the dollar as the global currency that I can only assume as a non-economist that such people know that what they are saying is inaccurate.

Because the US dollar is used, for instance, when Europe wants to buy oil from the Middle East, there is a constant outflow of dollars from the US. In terms of the US trade deficit, we see that the US exports its currency to the rest of the world, which has to offer products and resources in return. This is an excellent position for the US to be in as long as this house of cards does not collapse. If the world ever stops trading primarily in dollars, for instance by switching to euros, a large chunk of dollars will be offered back to the US. Unbeknown to many in the US, Saddam Hussein tried to do this very thing by making the euro the basis for payments for oil from Iraq just before the US invaded the country. In Europe, this switch to "petroeuros" was seen as one of the reasons behind the preemptive US strike on Iraq. (Two other countries considering a switch to petroeuros are Venezuela and Iran.) If the world ever decides to hand the dollars back to the US and ask for products and resources in return, everyone might be surprised to discover that there were too many US dollars in circulation and not enough economy to back them up.

A full discussion of this issue would go beyond the context of this book. Suffice it to say that if the US increased its efficiency and reduced oil imports drastically, even by switching to more expensive alternative fuels such as ethanol and biodiesel, the money would circulate within the domestic economy, creating jobs at home. Indeed, in Germany the effect on the job market of subsidies for renewable energy is positive. German politicians are not afraid of making the country uncompetitively expensive by supporting renewables. Rather, Germans understand that they are swapping resource imports for domestic jobs. Macroeconomists in the US should be following the example of German macroeconomists, including some of the most prominent ones, such as those at Deutsche Bank, which has released several reports in the past few years supporting biofuels, environmental taxation, and such.[7]

But switching to new technologies requires upfront investments. Some technologies, such as energy-saving lamps, cost only few dollars; others, such as Zero Energy Homes, may require that you pay the equivalent of, say, ten years of utility bills upfront. And in the middle we have energy-efficient cars, which pay for themselves after several years.

Is energy efficiency only for the rich? If so, the problem would concern not only consumers but also the ailing public sector. Fortunately, in the past few decades environmentalists and economists have found common ground. People from such diverse camps as Christie Whitman, former head of the Environmental Protection Agency under George W. Bush, and Robert F. Kennedy, son of the late Senator Robert Kennedy and one of the greatest critics of the Bush administration's environmental policies, agree on one thing: economics and ecology go very well together. Price signals on the market can be used to change consumption patterns.

Conservation Investments

You can even make a living from showing people how to save energy. Some investments in energy efficiency pay for themselves very quickly and even provide a considerable return on investment while others simply lower the total cost of ownership in the long term. How can you figure out where these returns are greatest and how can you make such investments if you do not have the funding? In recent years, the business of "energy contracting" has been booming. Experts show people where investments in efficient devices and equipment (such as heaters, insulation, and appliances) will make the most impact. In addition, these experts provide increased transparency in the itemization of energy costs.

Take for example the project at the Staudinger School in Freiburg, Germany.[8] At the end of the 1990s, around 282,000 euros were invested in the modernization of lighting equipment, the heating system, and water lines. Originally, the goal was to lower the consumption of electricity by 20 percent, heat by 30 percent, and water by 36 percent. In the end, the savings amounted to 30 percent for both electricity and heat and 70 percent for water. Private investors — most of them teachers, parents, and citizens of the community — made the initial investment because the city had no money for the project. An agreement was reached that the investors would get no more than a 6 percent return on their investment per annum, with any additional savings being paid back to the school and hence to the city budget. The director of the project, Dieter Seifried, estimates that the City of Freiburg will save up to one million euros if all the new equipment stays in service for the expected 18 years. In the first five years, the Staudinger project saved more than 400,000 euros in electricity, heating, and water costs. Clearly, the project will easily repay the initial investment plus the maximum 6 percent per annum. In addition, to reward children and teachers for their energy consciousness, the school gets 10,000 euros a year directly from the project.

What works in a German school can work across international borders: investors from industrialized countries can invest in the modernization of energy systems in developing countries. Western investors provide the equipment, and their compensation is based on the difference between the old and new levels of consumption.

So where did Seifried get this great idea? When I asked him, he answered: "Amory Lovins." Lovins, who heads the Rocky Mountain Institute, has been successfully promoting the idea of negawatts and least-cost planning in the US for over two decades.[9] It seems that Germans are quite willing to adopt best practices from other countries.

The concept of least-cost planning is quite simple: you ask yourself whether it would cost more to construct a new power plant or to invest in efficiency. Since the mid-1990s, this concept has truly taken off in projects in schools throughout the EU. In Germany, there is a project called EnergieSchule in the state of North-Rhein/Westfalia,[10] while Heidelberg is home to the E-Team project[11] and Hamburg has the 50-50 project.[12] The name of the project in Hamburg reflects the way savings are split, half of them going back to the school and the other half to the investors. In many cases, such projects cover everything from electricity, heating, and water savings to waste reduction. Currently more than 600 schools in Germany are taking part in such projects. In Great Britain, the Eco Schools campaign has been underway since 1994[13] and has expanded into other countries.

Schools are a good place for such campaigns for several reasons. First, schools generally do not have much money. Second (and more interesting), young people

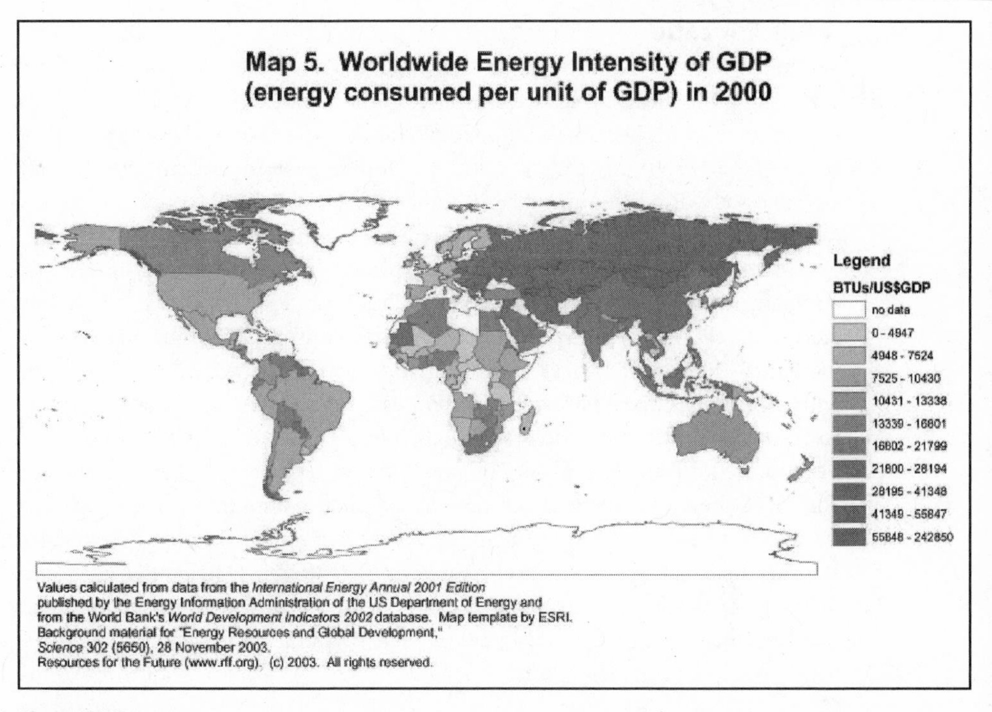

Figure 14.3:
Worldwide energy intensity (energy consumed per unit of GDP) in 2000. The United States is often criticized for having such high energy consumption — the second-highest per capita in the world behind Canada — but when it comes to "energy intensity" (energy consumption relative to gross national product), the US does not perform so poorly. Although Americans consume about twice as much energy per capita as Germans do, Americans at least get more added value from their energy than many developing nations, such as India or China. The worst major performer is the former Soviet Union, so if you have some extra cash to invest in efficiency, the greatest gains can be made there. Many countries lack funds to make such investments themselves, but energy contracting allows foreign investors to transfer energy-efficient technology across international borders. (Source: Resources for the Future, <www.rff.org>)

are more open to new ideas and like to do everything better than their parents. School children act as multipliers when they learn all about how to save energy at school and then pester their parents at home to do the same. The younger generation teaches the older generation to be aware of energy consumption.

It thus comes as no surprise that the Display Project of the French energy agency Energie-Cités, [14] which originally targeted municipal administrative buildings, has recently been expanded to include schools. This project focuses on transparency: a poster is hung up prominently in the building to show how great consumption was at the beginning of the project and how much has been saved in the meantime. Adults often do not even know how much energy they consume.

Adult Education

Adults do not react quickly when a sales pitch sounds like preaching, but they do react quickly when something affects their pocketbooks. This is where deZem of Berlin comes in. Founded at the beginning of 2003, this company sells only energy transparency. Like the Display Project, deZem tries to raise awareness about consumption. But, unlike the Display Project, deZem does so not by means of posters but by using real-time displays on computer screens.

Georg Riegel, founder of deZem, explains the concept:[15] "Usually, each company department pays electricity costs based on floor space or other parameters that reflect general overhead, not specific consumption." In other words, individual departments have no incentive to save energy. So Riegel came up with a combination of hardware and software that allows individual employees to track their own consumption and change their behavior. "Power costs can be allocated to a small group of people, cost awareness is very high, and employees begin thinking about turning off their computer screen when they go to lunch instead of letting it run with a screensaver on." And once this first step has been taken, people are open to more sweeping changes. Riegel estimates that by using his savings package companies large and small can reduce their electricity bill by up to 20 percent. The investments in his equipment pay for themselves in the short term.

Saving Energy as a Priority

It does not hurt much to forego a screensaver but what if you are really in love with that feeling you get when you put the pedal to the metal? Probably not much can be done overnight, but a new attitude first requires awareness about energy consumption. Transparency is the first step toward changing attitudes; higher prices are a step toward changing behavior.

The major drawback to this approach is that "higher prices" is not a hot seller, especially in the US. Some complain that environmentalists want us to live like cave dwellers. Obviously, that cannot be the goal. If you want to have a South African wine to go with your Argentinian steak, be my guest. A survey in Germany recently showed that people who voted Green have the most frequent-flyer miles;[16] Greens like to travel to all sorts of exotic places. Such mobility does not go well with the recent slogan of the Greens in Germany: "You are not stuck in a traffic jam; you are the traffic jam."

On the other hand, I would be willing to bet that most people do not even know how much electricity a 1,000-watt electric heater consumes if they let it run an hour, much less what this kilowatt-hour costs — or what resources were used by their utility to generate this electricity. Most people probably know only what they pay per month for power. Being aware of how much electricity we consume and when we consume it would at least allow us to change our behavior if we want to.

In addition, it would certainly help if more people understood that higher prices are not the same as higher costs. People can compensate for higher prices by lowering consumption. It is important to understand that, by lowering consumption, *costs* can be reduced even as *prices* increase. Transparency will allow us to determine the relationship between prices and costs.

We also can follow the following golden rule: muscle power, not fossil power. In the Netherlands, bicycles make up 27 percent of transportation. The figure in Germany is 12 percent. But it drops to 0.3 percent in the US, where urban infrastructure is designed for cars, not people. And if you think this car-based society is going to come back to haunt the US one day when oil becomes scarce, remember that Americans are already paying for burning fossil fuels instead of calories. The price is obesity.

Life Inside a Residential Power Plant

The US Department of Energy is developing a cost-competitive Zero Energy Home that will produce at least 90 percent of the energy its residents consume.[17] Such houses have moved beyond the pilot phase in Europe and are entering the mass market. While we still read of the first such homes going up in various parts of the US, entire neighborhoods of these houses are becoming net power exporters in Europe.

The pilot phase in Germany started back in the early 1990s. In 1992, the Fraunhofer Institute for Solar Energy Systems built the experimental Energy-Autonomous Solar House.[18] Aside from a few problems with fuel cells, the project was a grand success and led to the development of passive houses, which soak up so much solar energy passively that they almost always can do without heating, even in Germany.

While the Solar House was an expensive high-tech project, the passive houses that emerged from it are a wonderful combination of a little high-tech and a lot of low-tech, serving as a great reminder that we also use solar energy when we hang our clothes up to dry in the sun. Indeed, native peoples all over the world invented "solar houses" long ago, as Peter van Dresser reminded us way back in 1958 with his solar-heated adobe. And in the late 1970s, an entire residential area was constructed in Davis, California, focusing on energy conservation. Modern passive houses in central Europe have nearly all-glass south façades of special insulating windows that trap warmth inside even while letting a wide spectrum of sunlight in. In contrast, to reduce heat loss the northern sides generally have little window space.

This system is so effective — no drafts here — that a ventilation system is required to constantly refresh the inside air with outside air. Thus there is no danger of bacteria and fungi building up, which sometimes happens when indoor air is recirculated or (more common in Europe) when there is

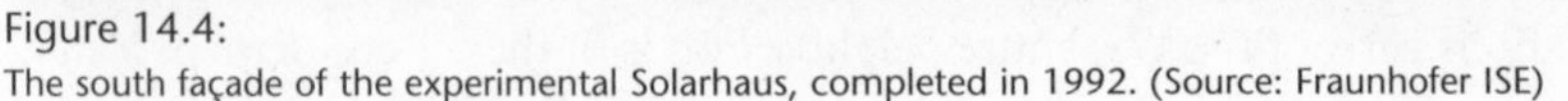

Figure 14.4:
The south façade of the experimental Solarhaus, completed in 1992. (Source: Fraunhofer ISE)

no ventilation at all. To prevent heat losses in the ventilation process, a heat exchanger is integrated into the system: the incoming air passes by the outgoing air in a coil and heats up as it enters. At the same time, the air is constantly filtered, making the air inside passive houses even better for those with allergies. The website of Germany's Passive House Institute[19] offers a more detailed overview of the principles behind passive houses.

The same ventilation system brings cooler air in at night during the summer. In addition, to prevent a house from overheating, the overhang of the south-side balconies is calculated to shade the windows underneath in the summer while letting in as much sunlight as possible in the winter. A little low-tech also helps: a deciduous tree can be planted south of the house to provide shade in summer; in winter, the barren tree allows the lower winter sun to shine through.

To qualify as a passive house, no high-tech equipment or material is required at all. Rather, Germany awards the label solely based on strict criteria for heating energy: no more than 1.5 liters of heating oil — the equivalent of 15 kilowatts — can be consumed per square meter (roughly 10 square feet) in one year, which would cost around 10 cents per square foot per year. A passive American house of some 2,200 square feet would thus have annual heating costs of around $220 rather than the current national average for a conventional house of about $1,000.

In Germany, the average old building uses up to 250 kilowatts per square meter per year for heating — more than 15 times as much as a passive house. In practice, values as low as 11 kilowatts per square meter per year have been reached in passive houses. Whether you reach those values using Styrofoam or recycled tire shreds as insulation does not matter as long as you get there.

And it pays to get there. While there are no nationwide criteria for what can be called a passive solar house in the US, if your house fulfills the heating requirements in Germany you are eligible for state funding. However, even without state aid the costs for passive houses have already proved to be competitive with those for conventional German houses. While passive houses require slightly higher investments up front, in the long run they are even cheaper than conventional houses because of far lower heating costs. Recently, passive houses in Germany have been made of straw bales, an excellent insulator and especially inexpensive since straw is a renewable waste product of agriculture.

Even on days far below zero, room temperatures in the 70s are nothing unusual in passive houses as long as the sun is shining, even without the assistance of the small emergency heater provided just for good measure. More problematic are those cold, gray days. But even then you don't have to resort to the heater; just invite some friends over. After all, a human gives off roughly as much heat as a conventional 100-watt light bulb.

From Passive to Active

The first residential area in Germany to consist solely of passive houses was built in the northern city of Hanover in the CEPHEUS project of 1999.[20] The obvious next step, now that power expenses had been reduced, was to implement another idea from the Solar House: photovoltaics integrated into roofs.

The result, a passive house with a PV roof, is a power plant you live in. An entire neighborhood of such homes, which produce more energy over the course of a year than their inhabitants consume, has gone up at the foot of the Black Forest in Freiburg's Solar Community.[21] Germany's 100,000 Roofs project[22] created a boom in the PV market around 2000. By 2002, 30 percent of the world's new installations were in Germany, making it a world leader in photovoltaics along with Japan, which had a similar program (unofficially called the 70,000 Roofs program).[23] Without excess power to sell to the grid, these houses would be unaffordable for many since the equivalent of a decade or so of heating and power costs have to be paid upfront. Only after 10 to 15 years do these homeowners break even. But in the decades afterward they will save a tidy sum by selling power to the grid instead of paying a power and gas bill.

Freiburg's Solar Community goes even further down the sustainability path. It is a car-free zone, with car sharing available at the communal parking

Figure 14.5:
The solar house of the future is available in serial production today — in Europe. Here is the Övolution® turnkey residential power plant. (Source: WeberHaus)

garage (with its PV roof), and it recycles its wastewater. Similar projects are appearing elsewhere in Europe, for instance Sweden's Bo01 City of Tomorrow project,[24] where not only building-integrated PV but also wind power will be generated in the car-free mini-city. In addition, the town's organic waste will be used in local agriculture. Other projects worth mentioning are Oslo's Pilestredet Park,[25] Helsinki's Att Viikki Sunh,[26] and Copenhagen's Agenda 21[27] programs. Because of colder temperatures and weaker solar radiation there, these Scandinavian houses do not meet the German passive house standard but heating costs generally still are reduced by 90 to 95 percent. And this housing is by no means limited to the rich. Because the comparatively homogeneous countries of northern Europe have great social cohesion, it should come as no surprise that, for instance, a group of 50 social apartments in the Nieuwland section of Amersfoort, Netherlands, are passive houses.[28] Indeed, social housing is also the target of an EU-wide program for building-integrated solar power called Renewable Energy Strategies for European Towns or RESET.[29] For a good overview of international building-integrated solar projects, visit the International Energy Agency's database.[30]

Along less affordable lines, the architect behind Freiburg's Solar Community has developed a house that turns on its axis, powered only by a

Figure 14.6:
The Heliotrop® takes up very little ground, thus allowing for excellent rainwater seepage.
(Source: Architect Rolf Disch)

100-watt motor, to follow the sun like a sunflower.[31] When heat is needed, the house tracks the sun; when the house needs cooling, it faces away.

The idea of a rotating house to track the sun is not new. French architect Patrick Marsilli developed the round rotating Domespace house in the 1980s.[32] Made entirely of wood, using cork as insulation, it won the German Environmental Prize in 1994.

Conservation in Renovation

All these ideas sound great, but we can't tear down all our cities overnight to put up car-free eco-villages. Luckily, there are ways of renovating houses to get them close to the zero-energy target even if centuries of building houses that don't face south prevent optimal energy saving.

The main principles of passive houses — heavy-duty insulation, insulating windows, a ventilation system with heat exchange for outdoor air, and possibly a heat pump — can be applied to just about any building. One of the most interesting options for renovations in cold regions is called "transparent insulation" (TI). TI consists of tiny, hollow, transparent tubes clad to the outside of the house that stand on their ends to trap solar heat. It operates on a principle similar to that of the hair of polar bears, which is transparent and hollow. The Nordic light heats up the air inside the polar bear's hair and shines through to the bear's black skin, which absorbs even more of the warmth. This method has proved so effective that one has to be careful not to overheat the house. Again, the solution to this problem is low-tech: a deciduous tree in front of the house will shade the TI just fine in summer and let the light through in winter.

Figure 14.7:
Here is a perfect example of high-tech meeting low-tech: the dark gray areas are TI, partially shaded here by a broad-leaf tree at the headquarters of the International Solar Energy Society in Germany. (Photo by the author)

In building or renovating their homes for energy efficiency, what can Americans learn from the European successes of the last ten years? While air conditioning accounts for about a fifth of the power consumption in the US, central and northern European homes rarely have air conditioners at all. Thus, in large parts of the US, passive houses will have to make air conditioners redundant. Luckily, increasing shading and using geothermal power to cool indoor air will take us a long way. However, poorly designed technology will not help. My parents have an energy-efficient air conditioning and hot water system at their new home in southern Mississippi but unfortunately the hot water tank is in the garage at the other end of the house from their shower. The water may be heated efficiently, but in the time it takes for the hot water to get to the master bathroom, I could be finished showering. Had the architect thought to put the hot water tank where the master bathroom is rather than where the cars are, the savings would be greater and hot water would be provided faster.

It would be a great step forward if we could begin to understand sustainability as a combination of new technology and common sense.

Chapter 15

Living to Learn

I found Finnish society beguiling on many levels, but in the end concluded that it could not serve as a blueprint for the United States. National differences matter. The Finns are special and so are we. Ours is a society driven by money, blessed by huge private philanthropy, cursed by endemic corruption and saddled with deep mistrust of government and other public institutions. Finns have none of those attributes.
—*Robert G. Kaiser, writing in* The Washington Post, *August 7, 2005, on what Americans can learn from Finland*

WHAT CAN AMERICANS LEARN FROM THIS BOOK? It seems that the main obstacles to a switch to renewables in the US are cultures and mindsets. Two of the people who have influenced European policy the most are Americans: former US president Jimmy Carter and Amory Lovins of the Rocky Mountain Institute. Quite possibly their ideas have influenced Europe more than the United States.

When I was looking for a US publisher for my book, every publishing house I contacted said that Americans would not be interested in a book on German energy policy. What did German policy have to do with the US? I was reminded of 1993, when politicians in the US Senate claimed that government health care coverage would be too expensive and ineffective. It turned out that in 1993 the US spent much more per capita on health care than other OECD countries, still could not cover some 15 percent of its population with insurance, and performed near the bottom of the OECD pack overall. The US was not learning from the best practices abroad. As the Rocky Mountain Institute writes, "In general, Europeans seem to understand better and employ more regularly the principles of integrated design, whole-systems thinking, smart growth, and life-cycle costing."[1] In other words, the first thing Americans will have to change is their minds.

Why is that? Part of the answer may be vested interests. The US is home to five of the seven largest oil companies in the world — known collectively as the

Seven Sisters — with BP and Shell being the only ones based in Europe (neither of them in Germany). While the US currently imports around 60 percent of the oil it consumes, this figure easily exceeds 90 percent for most European countries, a level the US probably will not reach for a generation. The relative lack of resources in Europe may explain why the EU is already looking for alternatives more intensively than the US is. If so, this explanation would also probably hold true for Japan.

Does that explain why the two European "Sisters" support renewables so much more than Chevron, Gulf, Texaco, and the now merged ExxonMobil? Is the general political climate in Europe simply more conducive to supporting renewables? Perhaps, but we should not overstate the success of Europe. While Europe has ratified the Kyoto Protocol, in all likelihood it will fall slightly short of its target. In addition, Europe certainly has its own vested interests. For example, as we saw, in numerous long-term road tests Greenpeace's SmILE car performed just as well as the Renault Twingo from which it was developed, but Renault has yet to implement Greenpeace's design changes, which would double the car's gas mileage without raising the price of the car considerably.

If we return to the list of obstacles to environmentalism in the US that Michael Shellenberger and Ted Nordhaus discuss in "The Death of Environmentalism," we see that they begin with "the radical right's control of all three branches of the US government." I take this statement not as an indictment of the Republican Party or an endorsement of the Democratic Party but simply as a sober observation that a group calling itself "neoconservative" (in a curious twist of terms, almost the exact opposite of "conservationist") now runs the US.

The distinction between Republicans and neoconservatives is crucial because, as Germany shows, conservatives can also be conservationists. At the same time, the differences between Democrats and Republicans should not be overstated. There are and always have been quite sensible Republicans, just as there are and always have been foolish Democrats. Then vice president Al Gore may have signed the Kyoto Protocol, but Democrats and Republicans alike opposed this move, with the US Senate voting 95-0 to reject it in the Byrd-Hagel Resolution, named partly after Democrat Robert Byrd, and with future Democratic presidential candidate John Kerry also voting against Kyoto.

The proclamation of the "death of environmentalism" seems unfair if we understand it as a criticism of environmentalists. The message that environmentalists must convey is complex, and part of their political opposition misinforms the public. Perhaps the greatest difference between public debate in Europe and in the US these days is that debate in the US is more ideological, less civil, and more mendacious. The course American environmentalists should take therefore has to be at least somewhat different than in Europe and certainly different from the manipulative neocon approach. We must be open and respectful, not misleading. Are we out to make enemies or friends?

If we believe (as polls suggest) that a wide majority of Americans support our agenda, then we should not be picky about our bedfellows. The country may still be split on gay marriage, abortion, gun control, and any number of other currently divisive issues, but we all agree that a clean environment, low unemployment, and energy independence are important. Our goal, then, is to demonstrate that a switch to renewables will provide all of these. A sustainable world will be a better world. In addition to adopting new technologies we will have to change our lifestyles, but the tradeoff will be positive. In October 2005, *The Wall Street Journal* reported that a Harris Interactive Poll had found that three-quarters of Americans believe that environmental protection is important and that "standards cannot be too high."[2] With such a majority, how can we be losing the debate?

Let us engage our fellow Americans in long, detailed, calm discussions. The more civil our discussions are, the more easily our complex arguments can be presented. We may not convince everyone, but the unconvinced may be reduced to those behind <www.raptureready.com>, who seem to believe that the faster we consume the planet, the faster Christ will come again.

Many societies have perished either because they refused to learn how to change or because their demise was seen as an inevitable act of God. Author Jared Diamond has described how the Vikings failed to learn from the native Greenlanders how to survive there once the warm period came to a close in the 15th century, and how the first Easter Island culture continued to cut down trees to transport the large stone monuments to their gods even when it was clear that the island's ecosystem would collapse without trees.[3]

In the decades ahead, the US could send its military all over the world to secure its supply of ever scarcer resources. Some Canadians fear that under Section 605 of NAFTA the US will rob Canada of its fossil riches. When will the US realize that it has more than enough renewable resources at home? The outcome is crucial not only for the US but also for the rest of the world. If its unique combination of neoconservatism and consumerism causes US society to collapse, it may take the whole planet down with it.

Jimmy Carter said it well at the end of the 1970s: "We must not be selfish or timid if we hope to have a decent world for our children and grandchildren."[4] Or as Article 20a of the German constitution puts it, "Mindful also of its responsibility toward future generations, the state shall protect the natural bases of life."[5]

Notes

Chapter 1

1. The Breakthrough Institute, www.thebreakthrough.org/images/Death_of_Environmentalism.pdf, cited Feb 6, 2006
2. Sharon Beder. *Power Play: The Fight to Control the World's Electricity*. W. W. Norton & Company, 2003
3. Ernst von Weizacker, Amory B. Lovins, and L. Hunter Lovins. *Factor Four: Doubling Wealth, Halving Resource Use — A Report to the Club of Rome.* Kogan Page, 1998
4. EnergyPulse, www.energypulse.net/centers/article/article_print.cfm?a_id=1065, cited Feb 6, 2006
5. "Bright Side of Blackouts," www.redherring.com/Article.aspx?a=13161&hed=The+Bright+Side+of+Blackouts, cited Feb. 6, 2006
6. "Berkeley Lab Study Estimates $80 Billion Annual Cost of Power Interruptions," www.lbl.gov/Science-Articles/Archive/EETD-power-interruptions.html, cited Feb. 6, 2006
7. "85 % der Deutschen befürworten Förderung Erneuerbarer Energien," www.energieportal24.de/artikel_1072.htm, cited Feb 6, 2006

Chapter 2

1. Herman Melville. *Moby Dick*. 1851, chapter cv
2. See *Nature*. May 15, 2003, p. 280
3. See the PDFs at www.millenniumassessment.org/en/products.aspx, cited Feb 8, 2006
4. See www.nhtsa.gov/cars/rules/CAFÉ/overview.htm, cited Feb. 9, 2006
5. See http://glogov.org/upload/public%20files/pdf/publications/working%20papers/workingpaper5.pdf, cited Feb 8, 2006
6. Green Budget Germany, www.eco-tax.info/downloads/GBGMemorandum2004.pdf, cited Feb 9, 2006
7. See www.nadeshda.org/foren/cl.politik.atom-presseschau/1464/, cited Feb 8, 2006

8. See www.grist.org/news/maindish/2004/07/13/griscom-kennedy/, cited Feb 8, 2006
9. Patricia H. Werhane. *Adam Smith and His Legacy for Modern Capitalism*. Oxford University Press, 1991
10. Herman Daly. *Beyond Growth: The Economics of Sustainable Development*. Beacon Press,1997
11. Donella H. Meadows et al. *The Limits to Growth: A Report.for the Club of Rome's Project on the Predicament of Mankind*. Earth Island, 1972 (for the 30-year update, see note 16)
12. Björn Lomborg. *Skeptical Environmentalist*. Cambridge University Press, 2001
13. See his "The State of Humanity: Steadily Improving" at www.cato.org/pubs/policy_report/pr-so-js.html, cited Feb 8, 2006
14. See "George Washington and I Are Subjects of Groundless Rumors" at www.sustainabilityinstitute.org/dhm_archive/index.php?display_article=vn526presstruthed, cited Feb 8, 2006
15. See "We told you so..." at www.heise.de/tp/r4/artikel/2/2485/1.html, cited Feb 8, 2006
16. Donella H. Meadows, Jorgen Randers, and Dennis L. Meadows. *Limits to Growth: The 30-Year Update*. Chelsea Green Publishing Company, 2004
17. See www.decroissance.org/ and "Pour une société de décroissance" at www.monde-diplomatique.fr/2003/11/LATOUCHE/10651, translated as "The World Downscaled" at http://mondediplo.com/2003/12/17growth, all cited Feb 8, 2006
18. See "From Here to Economy" at www.alternet.org/story/18518/, cited Feb 8, 2006
19. See www.worldwatch.org/pubs/vs/2005/, cited Feb 8, 2006
20. See www.worldwatch.org/features/vsow/2005/07/18/, cited Feb 8, 2006
21. *The Future of Nuclear Power: An Interdisciplinary MIT Study*. Massachusetts Institute of Technology, 2003
22. See "Uranium prices are set to climb" at www.energybulletin.net/4026.html, cited Feb 8, 2006, originally published in *International Herald Tribune*, Jan 15 2005
23. See Wir haben 30 Jahre verloren at www.zeit.de/2004/02/Meadows_Interview, Dec 31, 2003, cited Feb 8, 2006
24. See http://whitepaper.ises.org/ISES-WP-72.pdf, cited Feb 8, 2006
25. Hermann Scheer. *Solar Manifesto*. James & James Science Publishers, 2nd ed,, 2005
26. See "Is the world's oil running out fast?" at http://news.bbc.co.uk/1/hi/business/3777413.stm, cited Feb 8, 2006
27. See "Energy prospects after the petroleum age" at www.dbresearch.de/PROD/DBR_INTERNET_DE-PROD/PROD0000000000181487.PDF, cited Feb 8, 2006

Chapter 3

1. See www.worldenergy.org/wec-geis/publications/reports/etwan/policy_actions/chap_5_action2.asp, cited Feb 9, 2006

2. See "World Energy Outlook 2002," http://eneken.ieej.or.jp/en/seminar/021022p0002e01.pdf, cited Feb 9, 2006
3. See BP's *Statistical Review of World Energy*, www.bp.com/genericsection.do?categoryId=92&contentId=7005893, cited Feb 9, 2006
4. See "Philippines Ill-Prepared for Global 'Peak Oil'," www.bulatlat.com/news/5-34/5-34-oil.htm, cited Feb 9, 2006; other estimates state that we consume around four times more than we discover each year — see "Heinberg discusses 'peak oil' crisis," www.energybulletin.net/3283.html, cited Feb 9, 2006
5. See www.msnbc.msn.com/id/7559029/page/5/, cited Feb 9, 2006
6. See "White House: Saudi Arabia Producing Oil At 'Near Capacity'." Dow Jones Newswires, Apr 20, 2005
7. See www.gulfnews.com/Articles/BusinessNF.asp?ArticleID=155972, cited Feb 9, 2006
8. See "Time to think out of the barrel." *Baltimore Sun*. Apr 6, 2005
9. See "Saudis see no easing of oil prices." *Financial Times*. Feb 24, 2005
10. See "OPEC Struggles to Contain Oil Price Boom," Reuters, Mar 15, 2005
11. See www.msnbc.msn.com/id/7190109/, cited Feb 9, 2006
12. See www.iea.org/bookshop/add.aspx?id=197, cited Feb 9, 2006
13. See www.countercurrents.org/po-leggett250405.htm, cited Feb 9, 2006
14. See "IMF warns on risk of 'permanent oil shock'." *Financial Times*. Apr 7, 2005
15. See www.energybulletin.net/3230.html, cited Feb 9, 2006
16. See http://321energy.com/editorials/rubin/rubin041505.html, cited Feb 9, 2006
17. See "Goldman Sachs: Oil Could Spike to $105," Reuters, Mar 31, 2005
18. See "Will oil strike $380 a barrel by 2015?" at http://english.aljazeera.net/NR/exeres/73CE8286-740C-482B-8150-DA57696BC02F.htm, cited Feb 9, 2006
19. See www.washingtonpost.com/wp-dyn/articles/A63958-2005Apr18.html, cited Feb 9, 2006
20. See "Coal in a nice shade of green." *The New York Times*. Mar 25, 2005, p. A17
21. Matthew Simmons. *Twilight in the Desert: The Coming Saudi Oil Shock and the World Economy*. Wiley, 2005
22. See http://corporate.bmo.com/HarrisNesbitt/bresource/basicpoint/default.asp?id=4887, cited Feb 9, 2006
23. See www.pppl.gov/publications/pics/Oil_Peaking_1205.pdf, cited Feb 9, 2006
24. See note 12
25. See www.monbiot.com/archives/2004/08/23/living-with-the-age-of-entropy/, cited Feb 9, 2006
26. See note 3
27. See www.energybulletin.net/2544.html, cited Feb 9, 2006
28. See www.planetark.org/dailynewsstory.cfm/newsid/20367/story.htm, cited Feb 9, 2006

29. See www.nrdc.org/land/wilderness/artech/farcjobs.asp, cited Feb 9, 2006
30. Cited in Daniel Yergin. *The Prize*. Free Press, 1993
31. See www.eia.doe.gov/neic/press/press183.html, cited Feb 9, 2006
32. Paul Roberts. *The End of Oil*. Mariner Books, 2005
33. Amory B. Lovins. *Winning the Oil Endgame: Innovation for Profit, Jobs and Security*. Gardners Books, 2005
34. See www.rmi.org/images/other/Trans/T94-29_ReinventWheels.pdf, cited Feb 9, 2006, originally published in *Atlantic Monthly*, Jan 1995
35. See note 32
36. See "dena-Netzstudie" at www.deutsche-energie-agentur.de/page/index.php?id=2730&type=5, cited Feb 9, 2006
37. See note 32
38. See www.exxonmobil.de/unternehmen/service/publikationen/downloads/files/oeldorado2003.pdf, cited Feb 9, 2006
39. Personal communication
40. Colin J. Campbell, Frauke Liesenborghs, Jörg Schindler, and Werner Zittel. *Ölwechsel!* dtv, 2002
41. See note 21
42. See www.guardian.co.uk/life/feature/story/0,13026,1464050,00.html, cited Feb 9, 2006
43. See "Proving Proven Reserves Are Proven: An Art Form Or A Science?" at www.simmonsco-intl.com/files/SPE%20GCS%20Reservoir%20Study%20Group.pdf, cited Feb 9, 2006
44. See www.sec.gov/divisions/corpfin/forms/regsx.htm#gas, Reg. 210.4-10, cited Feb 9, 2006
45. See note 31
46. See note 38
47. See "War-Wary Saudis Move to Increase Oil Market Clout." *Washington Post*. Nov 30, 2002
48. See www.siliconinvestor.com/readmsg.aspx?msgid=21732172, cited Feb 6, 2006

Chapter 4

1. See www.sparksdata.co.uk/refocus/frames.asp?docid=32609195&accnum=1, cited Feb 10, 2006
2. See http://allafrica.com/stories/200403020625.html, cited Feb 10, 2006
3. See "Traditional Biomass Energy: Improving Its Use and Moving to Modern Energy Use," Jan 2004, www.renewables2004.de/doc/DocCenter/TBP11-biomass.pdf, cited Feb 10, 2006
4. See http://europa.eu.int/eur-lex/pri/en/oj/dat/2003/l_123/l_12320030517en00420046.pdf, cited Feb 10, 2006
5. See www.worldenergy.org/wec-geis/news_events/news/pressreleases/pr090904ser.asp, cited Feb 10, 2006
6. See "Bottom of the barrel," Dec 2, 2003 at www.guardian.co.uk/oil/story/0,11319,1097672,00.html, cited Feb 10, 2006
7. See www.european-climate-forum.net/events/norwich2003/pdf/ecf_norwich_woods.pdf, cited Mar 14, 2006

8. See www.chem.uu.nl/nws/www/research/e&e/biomas_a2004.htm, cited Feb 10, 2006
9. See "Bioenergie: Nachwuchs für Deutschland," www.oeko.de/service/bio/dateien/de/oemi_artikel_biomasse.pdf, cited Feb 10, 2006
10. See "Ecologically Optimized Extension of Renewable Energy Utilization in Germany," www.bmu.de/files/nutzung_ee_eng.pdf, cited Feb 10, 2006
11. See "Das unterschätzte Potential der Biomasse und deren Rolle im künftigen Energiemix," www.hermann-scheer.de/pdf/Energie_und_Management.pdf, cited Feb 10, 2006
12. See www.unep-wcmc.org/index.html?http://www.unep-wcmc.org/forest/protection_statistics.htm~main, cited Feb 10, 2006
13. See www.energienetz.de/pre_cat_43-id_111-subid_535.html, cited Feb 10, 2006
14. See "The Logs of War: How Timber Fuels the World's Worst Conflicts," Jan 2004, http://MondeDiplo.com/2004/01/15timber, cited Feb 10, 2006
15. See www.fao.org/documents/show_cdr.asp?url_file=/docrep/007/ae153e/AE153e04.htm, cited Feb 10, 2006
16. See note 15
17. See http://verwertung.dm1-2.de/page/main.php?Lang=English&Menu=4&SubMenu=4&SubMenu2=0&Page=, cited Feb 10, 2006
18. Available at www.oilcrisis.com/cleveland/OilAndCulture.pdf, cited Feb 10, 2006
19. See note 9
20. See www.fairbiotrade.org/, cited Feb 10, 2006
21. This plant has not been widely tested in North America; see the Oak Ridge National Laboratory's assessment at http://bioenergy.ornl.gov/papers/miscanthus/miscanthus.html, cited Feb 24, 2006
22. See www.udo-leuschner.de/rezensionen/rc9209alt.htm, cited Feb 10, 2006
23. See www.solarbuch.de/html/content.htm, cited Feb 10, 2006
24. See "Evaluating Environmental Consequences of Producing Herbaceous Crops For Bioenergy." *Biomass and Bioenergy*. Vol. 14, No. 4 (1998), pp. 317-324 at www.public.iastate.edu/~brummer/papers/McLaughlin-Walsh.pdf, cited Feb 10, 2006
25. See www.fnr.de/, cited Feb 10, 2006
26. See "Widescale Biodiesel Production from Algae," www.unh.edu/p2/biodiesel/article_alge.html, cited Feb 10, 2006
27. See www.choren.com/de/, cited Feb 10, 2006
28. See follow-up from Jul 2004 at www.discover.com/issues/jul-04/features/anything-into-oil/, cited Feb 10, 2006
29. See www.fibl.org/, cited Feb 10, 2006
30. Richard Manning. *Against the Grain: How Agriculture Has Hijacked Civilization*. North Point Press, 2004; see also his "The Oil We Eat" at www.harpers.org/TheOilWeEat.html, cited Feb 10, 2006

Chapter 5

1. Italo Calvino. *The Baron in the Trees*. Harvest Books, 1977
2. Clive Ponting. *Green History of the World*. Penguin, 1993

3. Barbara Freese. *Coal: A Human History.* Perseus Books Group, 2003
4. John Evelyn. *Fumifugium.* Out of print, 1661
5. Robert Hasenclever. *Über die Beschädigung der Vegetation durch saure Gase.* Out of print, 1876
6. Mark Hertsgaard. *Earth Odyssey.* Broadway, 1999
7. See "A great wall of waste." *The Economist.* Aug 19, 2004, www.economist.com/PrinterFriendly.cfm?Story_ID=3104453, cited Feb 11, 2006
8. See "Modern China is facing an ecological crisis." *Financial Times.* Jul 26, 2004
9. See www.fossil.energy.gov/programs/powersystems/futuregen/, cited Feb 11, 2006
10. See www.issues.org/issues/20.2/john.html, cited Feb 11, 2006
11. See "Putting the World on a Low Carbon Diet." *Time Magazine.* May 10, 2004, www.time.com/time/europe/next/040517/lowcarbon.html, cited Feb 11, 2006
12. See "North Sea burial for greenhouse gases." *The Guardian.* Aug 1, 2004, http://politics.guardian.co.uk/print/0,3858,4983418-107983,00.html, cited Feb 11, 2006
13. See "2050 sind CO2-Speicher voll." *Neues Deutschland.* www.nd-online.de/artikel.asp?AID=58031&IDC=9&DB=Archiv, cited Feb 11, 2006
14. See www.g-o.de/index.php?cmd=focus_detail2&f_id=147&rang=6, cited Feb 11, 2006
15. See www.npg.org/forum_series/fall04forum.html, cited Feb 11, 2006
16. See www.dti.gov.uk/energy/environment/eerp/reports/ps_001.pdf, cited Feb 11, 2006
17. See "Greenhouse gas buildup seen as risk to oceans," www.boston.com/news/nation/articles/2004/07/16/greenhouse_gas_buildup_seen_as_risk_to_oceans?mode=PF, cited Feb 11, 2006
18. Jeffrey Michel. *Status and Impacts of the German Lignite Industry.* www.acidrain.org/pages/publications/reports/APC18.pdf, cited Feb 11, 2006
19. Interview on Sep 21, 2004
20. See "A Medieval German Hamlet Keeps the Bulldozers at Bay." *The New York Times.* Jul 27, 2004, www.heuersdorf.de/NYT27072004.html, cited Feb 11, 2006
21. See www.rspb.org.uk/scotland/action/disaster/index.asp, cited Feb 11, 2006
22. See http://cta.policy.net/, cited Feb 11, 2006
23. See Chapter 6, note 28
24. See www.clean-energy.us/success/great_plains.htm, cited Feb 11, 2006
25. See www.clean-energy.us/projects/eastman_power_magazine.htm, cited Feb 11, 2006
26. See www.innovationsreport.de/html/berichte/energie_elektrotechnik/bericht-32620.html, cited Feb 12, 2006
27. See www.memagazine.org/backissues/dec03/features/coalcell/coalcell.html, cited Feb 12, 2006
28. The German text speaks of "eine Energie, die die Brücke sein kann zwischen dem Heute und der Zukunft aus Wind, Wasser und Sonne," see www.dskag.de/medien/pdf/T-1097770604.pdf, cited Feb 12, 2006

29. See www.businessweek.com/print/bwdaily/dnflash/aug2004/nf20040823_9499_db_81.htm?db, cited Feb 12, 2006
30. See www.solarstromag.net/, cited Feb 12, 2006
31. See www.mpimet.mpg.de/index.html, cited Feb 12, 2006
32. See www.abc.net.au/rn/science/earth/stories/s729902.htm, cited Feb 12, 2006
33. See www.worldwatch.org/live/discussion/78/, cited Feb 12, 2006
34. See "Lessons & Limits of Climate History: Was the 20th Century Climate Unusual?," www.marshall.org/article.php?id=136, cited Feb 12, 2006
35. See www.sepp.org/, cited Feb 12, 2006
36. See www.peakoil.org and http://dieoff.org/, cited Feb 12, 2006
37. See Chapter 2, note 12
38. See www.enn.com/news/2003-06-19/s_5241.asp, cited Jul 11, 2003
39. See www.heise.de/tp/r4/artikel/20/20990/1.html, cited Feb 12, 2006

Chapter 6

1. See www.umweltinstitut.org/frames/all/m410.htm, cited Feb 12, 2006
2. See *Dilemma Study: Study of the Contribution of Nuclear Power to the Reduction of Carbon Dioxide Emissions from Electricity Generation,* http://europa.eu.int/comm/energy_transport/library/dilemma.pdf, cited Feb 13, 2006
3. See www.npg.org/forum_series/fall04forum.html, cited Feb 13, 2006
4. See www.commondreams.org/views05/0415-23.htm, cited Feb 13, 2006
5. See http://web.mit.edu/nuclearpower/, cited Feb 13, 2006
6. See www.greenpeace.org/international/news/no-to-new-nukes-go-wind, cited Feb 13, 2006
7. See www.jbsenergy.com/Energy/Papers/Rancho_Seco/rancho_seco.html, cited Feb 15, 2006
8. The original German reads: "Wir hätten das Kernkraftwerk Stade auch ohne die Vereinbarung mit der Bundesregierung stillgelegt." See http://de.biz.yahoo.com/031114/71/3r4os.html, cited Feb 23, 2004
9. See www.planetark.org/dailynewsstory.cfm/newsid/19626/story.htm, cited Feb 13, 2006
10. See www.iht.com/IHT/BJ/00/bj061500.html, cited Feb 23, 2004
11. See note 5
12. See www.planetark.org/dailynewsstory.cfm/newsid/25914/story.htm, cited Feb 13, 2006
13. The original German reads: "Die Investitionskosten für Atomkraftwerke sind so hoch, dass an einen Neubau gar nicht zu denken ist." See www.nadeshda.org/foren/cl.politik.umwelt/p1366s1373a20.html, cited Feb 13, 2006
14. See http://solstice.crest.org/repp_pubs/articles/envImp/06analysis.htm, cited Feb 23, 2004
15. See www.solarfabrik.de/, cited Feb 23, 2004
16. See http://europa.eu.int/comm/public_opinion/archives/eb/ebs_169.pdf, cited Feb 13, 2006
17. See note 5

18. See www.world-nuclear.org/, cited Feb 15, 2006
19. See www.wind-energie.de/index.php?id=166, cited Feb 17, 2006
20. See "ESSO Energieprognose 2003," www.esso.de/ueber_uns/info_service/publikationen/downloads/files/Energieprognose_03.pdf, cited Feb 15, 2006
21. Personal communication, Jan 29, 2004
22. See www.enn.com/news/2003-12-23/s_9206.asp, cited Mar 2, 2004
23. See "Generating Solutions: How States Are Putting Renewable Energy Into Action," www.uspirg.org/reports/generatingsolutions2_02.pdf, cited Feb 15, 2006
24. See "The present and future use of solar thermal energy as a primary source of energy," www.iea.org/textbase/papers/2005/solarthermal.pdf, cited Feb 15, 2006
25. See www.energyrichjapan.info/pdf/EnergyRichJapan_summary.pdf, cited Feb 15, 2006
26. See www.heise.de/tr/artikel/print/42207, cited Feb 17, 2006
27. See www.climatetechnology.gov/library/ 2003/tech-options/tech-options-1-3-2.pdf, cited Feb 15, 2006
28. See "Nuclear power is the only green solution," May 24, 2004, http://argument.independent.co.uk/commentators/story.jsp?story=524230, cited May 25, 2004. See also www.yesmagazine.org/article.asp?ID=1065, cited Feb 24, 2006
29. See note 28
30. See www.wind-works.org/articles/BreathLife.html, cited Feb 15, 2006
31. See "Nuclear Power 'Can't Stop Climate Change'," www.commondreams.org/headlines04/0626-05.htm, cited Feb 15, 2006
32. See "Level the Energy Playing Field," www.truthout.org/cgi-bin/artman/exec/view.cgi/9/4534, cited Feb 15, 2006
33. See www.japanfocus.org/article.asp?id=460, cited Feb 15, 2006
34. See "Can we really afford more nuclear power?" *Times Standard*, www.times-standard.com/cda/article/print/0,1674,127%7E2906%7E2342380,00.html, cited Jul 11, 2002
35. See www.ymp.gov/factsheets/doeymp0338.htm, cited Jul 11, 2002
36. See www.ymp.gov/new/earthquake_pr.htm, cited Jul 11, 2002
37. See "Senators Declare Support for Waste Site." *The New York Times.* Jul 9, 2002, p. A18
38. See http://enn.com/news/wire-stories/2002/06/06262002/ap_47665.asp, cited Jul 11, 2002
39. See www.yuccamountain.org/faq.htm, cited Jul 11, 2002
40. See www.ymp.gov/new/transbro/insidecover.htm, cited Jul 11, 2002
41. See www.nei.org/doc.asp?catnum=2&catid=243, cited Feb 15, 2006

Chapter 7

1. See www.foes-ev.de/news24/2artikel3.html, cited Feb 17, 2006
2. See http://europa.eu.int/comm/research/energy/pdf/key_messages_en.pdf, cited Feb 17, 2006
3. See "Bedeutung von Erdgas als neuer Kraftstoff im Kontext einer nachhaltigen

Energieversorgung," www.kunden.01dd.de/erdgas2004/infospalte/downloads/WI_Report_deu.pdf, cited Jun 29, 2004

4. See http://smarteconomy.typepad.com/smart_economy/2006/01/as_oil_gas_pric.html, cited Feb 17, 2006
5. See www.newscientist.com/article.ns?id=mg18725124.500, cited Feb 17, 2006
6. See www.newstatesman.com/nssubsfilter.php3?newTemplate=NSArticle_Ideas&newDisplayURN=200405170018, cited Feb 17, 2006

Chapter 8

1. Figures taken from phone call with Gero Hollmann
2. Figures taken from phone call with Günther Schmarje
3. See www.sonnenstromag.de/, cited Feb 17, 2006
4. Ingo Bert Hagemann. *Gebäudeintegrierte Photovoltaik.* Müller Rudolf, 2002
5. See www.bsi-solar.de/marktdaten.asp, cited Feb 17, 2006
6. See www.prnewswire.co.uk/cgi/news/release?id=106805, cited Feb 17, 2006
7. Figures taken from email correspondence with Klaus Kiefer
8. The German original reads: "In vier bis fünf Jahren werden Solarzellen wirtschaftlich sein und dann folgt eine lange Wachstumsphase." See www.abendblatt.de/daten/2003/10/20/220165.html, cited Feb 17, 2006
9. See www.mondediplomatique.de/pm/2004/02/13.mondeText.artikel,a0060.idx,17, cited Feb 17, 2006
10. See "World energy, technology and climate policy outlook," http://europa.eu.int/comm/research/energy/pdf/weto_chapter4.pdf, cited Feb 17, 2006
11. See www.solarserver.de/news/news-1851.html, cited Feb 17, 2006
12. See Peter Fairley. "Solar on the Cheap." *Technology Review.* Jan-Feb 2002, pp. 48-53 at http://physics.ucsc.edu/~sacarter/Solar%20on%20the%20cheap.pdf, cited Feb 23, 2006
13. See www.nsf.gov/od/lpa/newsroom/pr.cfm?ni=73, cited Jun 7, 2004
14. See www.wissenschaft.de/wissen/news/232026, cited Feb 17, 2006
15. See www.cheresources.com/solarleaves.shtml, cited Feb 23, 2006
16. See www.synergypowercorp.com/proj9.htm, cited Feb 17, 2006
17. See www.sparksdata.co.uk/refocus/frames.asp?docid=59598941&accnum=1, cited Feb 17, 2006

Chapter 9

1. See http://service.spiegel.de/digas/servlet/epaper?Q=SP&JG=2004&AG=14&SE=1, cited Feb 23, 2006
2. See note 1
3. See www.skygeneration.ca/projects/enviroassessment.doc, cited Feb 23, 2006
4. See www.heise.de/tp/r4/artikel/20/20701/1.html, cited Feb 23, 2006
5. See www.solarserver.de/news/news-1732.html, cited Feb 23, 2006
6. See www.sueddeutsche.de/wirtschaft/artikel/68/27041/, cited Feb 23, 2006
7. See http://solarwirtschaft.solarinfo.de/magazin/print.cfm?artikel=/News/mag_1638.cfm&id=1729, cited Feb 23, 2006

8. See www.eeg-aktuell.de/index.php/article/articleview/116/1/2, cited Feb 23, 2006
9. See note 1
10. S. Orloff and A. Flannery. *Wind Turbine Effects on Avian Activity, Habitat Use, and Mortality in Altamont Pass and Solano County Wind Resource Areas, 1989-1991*. BioSystems Analysis, Inc., 1992
11. See "Bird Risk Behaviors and Fatalities at the Altamont Pass Wind Resource Area," www.nrel.gov/docs/fy04osti/33829.pdf, cited Feb 23, 2006
12. See www.nationalwind.org/pubs/avian_collisions.pdf, cited Mar 5, 2003
13. See note 12
14. Stream at www.wamu.org/ram/2001/r1010619.ram, cited Feb 23, 2006
15. Personal communication
16. The German reads, "Der BUND befürwortet den weiteren Ausbau der Windenergie-Nutzung in Deutschland als eine dezentrale erneuerbare Energiequelle. Dieser Ausbau muß in Natur und Mensch schonender und geordneter Weise erfolgen. Windenergie als eine besonders umweltfreundliche und dauerhafte Energiequelle wird bei der Stromversorgung im ökologischen Energie-Mix der nachhaltigen Energiewirtschaft eine wichtige Rolle spielen." See BUND's "Positionspapier" at www.bund.net/lab/reddot2/pdf/windenergie.pdf, cited Feb 23, 2006
17. See www.windpower.org/en/tour/env/birds.htm, cited Feb 23, 2006
18. See "Einfluss der Windenergienutzung auf die Avifauna im Binnenland" by Frank Bergen at www.wind-energie.de/fileadmin/dokumente/Themen_A-Z/Vogelschutz/Studie_diss_zusammenfassung.pdf, cited Feb 23, 2006
19. See www.innovations-report.com/html/profiles/profile-1105.html, cited Feb 23, 2006
20. See www.windpower.org/en/tour/design/quietma.htm, cited Feb 23, 2006
21. See www.eru.rl.ac.uk/web.htm, cited Feb 23, 2006
22. See www.denverpost.com/Stories/0,1413,36%257E33%257E1186542%257E,00.html, cited Mar 5, 2003
23. American industry experts and insiders point out that there are often disputes about patents, and that Kenetech was working on such a direct-drive system at the time, but that does not explain the apparent economic espionage via Echelon. See the details provided in issue 40/1999 of the most highly regarded German weekly *Die Zeit*, "Verrat unter Freunden: Wie die NSA, Amerikas größter und verschwiegenster Geheimdienst, deutsche Firmen ausspioniert und dabei einen Milliardenschaden anrichtet."

Chapter 10

1. German Bundestag. *Möglichkeiten geothermischer Stromerzeugung in Deutschland*. Arbeitsbericht Nr. 84, Feb 2003. See www.tab.fzk.de/de/projekt/zusammenfassung/ab84.pdf, cited Feb 23, 2006
2. See www.geo-energy.org/aboutGE/basics.asp, cited Feb 23, 2006
3. Telephone interview with Dagmar Oertel

Chapter 11

1. See http://news.bbc.co.uk/1/hi/sci/tech/4719334.stm, cited Feb 24, 2006

2. Quoted in *The Wall Street Journal.* http://online.wsj.com/article/0,,SB108431594329008644-search,00.html?collection=autowire%2F30day&vql_string=fuel+cells%3Cin%3E%28article%2Dbody%29. Norihiko Shirouzu and Jeffrey Ball. "Revolution Under the Hood: In Major Wave of Innovation, Car Makers Explore New Ways To Cut Fuel Use and Pollution." May 12, 2004, p. B1
3. See www.hydrogennow.org/HNews/PressArea/Carbon.htm, cited Feb 24, 2006
4. Jeremy Rifkin. *The Hydrogen Economy: The Creation of the Worldwide Energy Web and the Redistribution of Power on Earth.* Jeremy P. Tarcher, 2003
5. Stephen Eaves and James Eaves. "A Cost Comparison of Fuel-Cell and Battery Electric Vehicles." See www.modenergy.com/BEVs%20vs%20FCVs%20EavesEaves%20120603.pdf, cited Feb 24, 2006
6. Roel Hammerschlag and Patrick Mazza. *Carrying the Energy Future: Comparing Hydrogen and Electricity for Transmission, Storage and Transportation.* Institute for Lifecycle Environmental Assessment, 2004. See www.ilea.org/downloads/MazzaHammerschlag.pdf, cited Feb 24, 2006
7. See www2.warwick.ac.uk/newsandevents/pressreleases/NE100000009439/, cited Feb 24, 2006
8. See www.planetark.com/dailynewsstory.cfm/newsid/27514/story.htm, cited Feb 24, 2006
9. See www.aller-zeitung.de/az-lokal/227850.html, cited Feb 24, 2006
10. See www.guardian.co.uk/comment/story/0,,1097622,00.html, cited Feb 24, 2006
11. See dena's "Grid Study" at www.deutsche-energie-agentur.de/page/fileadmin/DeNA/dokumente/Programme/Kraftwerke_Netze/dena_Grid_Study_Summary_2005-03-23.pdf, cited Feb 24, 2006
12. See www.emagazine.com/view/?171, cited Feb 24, 2006
13. *The Hydrogen Economy: Opportunities, Costs, Barriers, and R&D Needs.* The National Academies Press, 2004. See http://books.nap.edu/catalog/10922.html, cited Feb 24, 2006
14. See www.vdi.de/vdi/news/index.php?ID=1014852, cited Feb 24, 2006
15. See www.mtu-online.com/cfc/en/cfcs/cfcs.htm, cited Feb 24, 2006
16. See www.innovationsreport.de/html/berichte/energie_elektrotechnik/bericht-32723.html, cited Feb 24, 2006
17. See www.wagingpeace.org/menu/resources/sunflower/2004/01_sunflower.htm#8a, cited Feb 24, 2006
18. See www.epa.gov/epaoswer/non-hw/muncpl/landfill/bioreactors.htm, cited Feb 24, 2006

Chapter 12

1. See his entry on "Wave and Tidal Energy Conversion" in Elsevier's *Encyclopedia of Energy,* Vol 6, 2004
2. See www.sfenvironment.com/articles_pr/2004/article/010004.htm, cited Feb 25, 2006
3. See www.marineturbines.com/home.htm, cited Feb 25, 2006
4. See www.engb.com/, cited Feb 26, 2006

5. See www.wavedragon.net/, cited Feb 26, 2006
6. See www.wavegen.co.uk/, cited Feb 26, 2006
7. See www.wave-energy.net/Projects/ProjDescriptions/pelamis.htm, cited Feb 26, 2006

Chapter 13

1. The original German reads: " Zur Aufrechterhaltung der Versorgungssicherheit [wird] eine so genannte Regelreserve benötigt. Dies bedeutet in der Praxis aber nichts anderes als herkömmliche Kraftwerke, die bei Windmangel den benötigten Strom produzieren. Diese Reserve muss, vor allem wenn die im Raum stehenden Ausbaupläne im Bereich der Windenergienutzung realisiert werden sollten, zunehmend durch im Mittellastbereich arbeitende Kraftwerke oder gar Grundlastkraftwerke übernommen werden. Nur so können Netzüberlastungen oder -zusammenbrüche vermieden werden. Diese Kraftwerke müssten jedoch sehr häufig 'angedrosselt', das heißt: mit reduzierter Leistung, laufen. Das bedeutet aber wiederum, dass sie nicht den optimalen Wirkungsgraderreichen." See www.sonnenseite.com/fp/archiv/Art-Umweltpolitik/3784.php, cited Sep 30, 2003
2. See www.encorp.com/content.asp?cmsID=41, cited Feb 26, 2006
3. See "DG: Coming to the Grid's Rescue" at www.forester.net/de_0501_dg.html, cited Feb 26, 2006
4. See "Economic Growth and the Central Generation Paradigm" at www.primaryenergy.com/articles/presentations/IAEE%20final%20article%20071404.pdf, cited Feb 26, 2006
5. See note 4
6. See "The choice not to buy: energy savings and policy alternatives for demand response" at www.allbusiness.com/periodicals/article/776152-1.html, cited Mar 12, 2006
7. See "Blackout Adds to Long List of Utilities' Woes." *The Baltimore Sun*. Aug 24, 2003
8. See www.ambientdevices.com/cat/orb/orborder.html, cited Mar 12, 2006
9. See www.grist.org/news/powers/2003/08/28/lock/, cited Feb 26, 2006
10. See www.hometoys.com/htinews/jun00/articles/aladn/aladn.htm, cited Feb 26, 2006
11. See www.osgi.org/, cited Feb 26, 2006
12. See www.homeplug.org/en/index.asp, cited Feb 26, 2006
13. See www.nchip.com/, cited Feb 26, 2006
14. See www.efr.de/de/index2.htm, cited Feb 26, 2006
15. See "Energiemanagement als Zukunftsmarkt" at www.heise.de/newsticker/meldung/42277, cited Feb 27, 2006
16. See www.dsireusa.org/library/includes/incentive2.cfm?Incentive_Code=PA06R&state=PA&CurrentPageID=1, cited Feb 27, 2006
17. See www.a0q.co.uk/Financing_Agreements_Secures_Financing_164.html, cited Feb 27, 2006
18. See www.epri-intelligrid.com/intelligrid/home.jsp, cited Feb 27, 2006
19. See www.economist.com/science/tq/displayStory.cfm?story_id=2476988, cited Feb 27, 2006

20. See www.otii.com/technology-products.html, cited Feb 27, 2006
21. See www.gridwise.org/, cited Feb 27, 2006
22. See www.dramcoalition.org/, cited Feb 27, 2006
23. See www.energy.ca.gov/pier/final_project_reports/CEC-500-2005-096.html, cited Feb 27, 2006
24. See Chapter 1, note 2
25. See www.siemens.com/index.jsp?sdc_p=t15cz3s5u20o1193107d1187140p FEn1193082flmi1193082&sdc_sid=20139682410&, cited Feb 27, 2006

Chapter 14

1. See www.energycentral.com/content/newsletter_creation/ energybizinsider_html_creation.cfm?articleid=71, cited Feb 28, 2006
2. Stream at www.wamu.org/programs/dr/03/05/20.php, cited Feb 28, 2006
3. See www.umweltbundesamt.de/uba-info-presse/2004/pd04-010.htm, cited Feb 28, 2006
4. See Chapter 1, note 3
5. See www.stcwa.org.au/journal/200805/1123914930_19114.html, cited Mar 2, 2006
6. Peter Hennicke and Dieter Seifried. *Das Einsparkraftwerk. Eingesparte Energie neu nutzen*. 2002, Hirzel
7. Its English reports are found at www.dbresearch.com/servlet/ reweb2.ReWEB?rwsite=DBR_INTERNET_EN-PROD&$rwframe=0, cited Feb 28, 2006
8. See www.eco-watt.de/proj_staud1.htm, cited Feb 28, 2006
9. See www.rmi.org/, cited Feb 28, 2006
10. See www.ea-nrw.de/_infopool/info_details.asp?InfoID=1401, cited Feb 28, 2006
11. See http://europa.eu.int/comm/environment/networks/doc/ heidelburg_eteams.pdf, cited Feb 28, 2006
12. See www.energie-cites.org/documents/stuttgart/w2_hamburg_2.pdf, cited Feb 28, 2006
13. See www.eco-schools.org.uk/, cited Feb 28, 2006
14. See www.energie-cites.org/, cited Feb 28, 2006
15. Personal correspondence
16. See "Lügen in Zeiten des Staus." *Die Zeit*. May 13, 2004 at www.zeit.de/2004/21/Verkehr, cited Feb 28, 2006
17. See www.fsec.ucf.edu/bldg/active/zeh/, cited Feb 28, 2006
18. See www.nrel.gov/ncpv/documents/germany.html, Feb 28, 2006
19. See the English at www.passiv.de/, cited Feb 28, 2006
20. See the English at www.cepheus.de/, cited Feb 28, 2006
21. See www.freiburgersolarfonds.de/default.asp?id=44, cited Feb 28, 2006
22. See www.germany-info.org/relaunch/info/publications/infocus/environment/ renew.html, cited Feb 28, 2006
23. See www.solarserver.de/solarmagazin/artikelseptember2001-e.html, cited Feb 28, 2006

24. See http://home.att.net/~amcnet/bo01.html, cited Mar 2, 2006
25. See www.skanska.com/index.aspx?id=386, cited Mar 2, 2006
26. See www.arrak.com/pages/att_viikki_sunh/att_viikki_sunh_e_.htm, cited Mar 2, 2006
27. See www.opet.dk/Agenda21/A21chp.htm, cited Mar 2, 2006
28. See www.bear.nl/content/bearafoort.html, cited Mar 2, 2006
29. See www.managenergy.net/products/R464.htm, cited Mar 2. 2006
30. See www.pvdatabase.com/search_form.cfm, cited Mar 2, 2006
31. See www.rolfdisch.de/project.asp?id=45&sid=1404685970, cited Mar 2, 2006
32. See www.domespace.com/vinter/, cited Mar 2, 2006

Chapter 15

1. See www.rmi.org/sitepages/pid907.php, cited Mar 2, 2006
2. See http://online.wsj.com/public/article/SB112914555511566939.html?mod=todays_free_feature, cited Mar 2, 2006
3. Jared Diamond. *Guns, Germs, and Steel: The Fates of Human Societies.* W. W. Norton & Company, 1999
4. See www.pbs.org/wgbh/amex/carter/filmmore/ps_energy.html, cited Mar 2, 2006
5. See www.bundesregierung.de/en/Federal-Government/Function-and-constitutional-ba-, 10221/II.-The-Federation-and-the-Lae.htm, cited Mar 2, 2006

Index

C

D

E

F

G

H

O

P

Y

Z

About the Author

BORN IN NEW ORLEANS and raised mostly in Mississippi, **Craig Morris** holds degrees from Tulane University and the University of Texas at Austin; he has also studied at universities in France and Germany and taught at universities in the US and Germany. Since 1998, he has been working full-time as a translator specializing in the energy sector and financials.

Since 2002, he has been covering environmental and energy issues and EU/US policies as a journalist, mostly in German. In his off-time, he is a passionate cyclist (recumbent trike, racing bike, mountain bike), a jazz singer who has been on tour in Europe and Asia, and a test-taster of exotic foods. He has two children (and counting) and does not wish to leave reproduction to those not worried about overpopulation.

If you have enjoyed *Energy Switch,* you might also enjoy other

Books to Build a New Society

Our books provide positive solutions for people who want to make a difference. We specialize in:

Sustainable Living • Ecological Design and Planning
Natural Building & Appropriate Technology • New Forestry
Environment and Justice • Conscientious Commerce
Progressive Leadership • Resistance and Community • Nonviolence
Educational and Parenting Resources

New Society Publishers

ENVIRONMENTAL BENEFITS STATEMENT

New Society Publishers has chosen to produce this book on recycled paper made with 100% post consumer waste, processed chlorine free, and old growth free.

For every 5,000 books printed, New Society saves the following resources:[1]

23	Trees
2,050	Pounds of Solid Waste
2,255	Gallons of Water
2,941	Kilowatt Hours of Electricity
3,726	Pounds of Greenhouse Gases
16	Pounds of HAPs, VOCs, and AOX Combined
6	Cubic Yards of Landfill Space

[1]Environmental benefits are calculated based on research done by the Environmental Defense Fund and other members of the Paper Task Force who study the environmental impacts of the paper industry.

For a full list of NSP's titles, please call 1-800-567-6772 *or check out our web site at:*

www.newsociety.com

New Society Publishers